Citrus

Citrus

A HISTORY

PIERRE LASZLO

The University of Chicago Press   Chicago and London

The University of Chicago Press, Chicago 60637
The University of Chicago Press, Ltd., London
© 2007 by The University of Chicago
All rights reserved. Published 2007
Paperback edition 2008
Printed in the United States of America

17 16 15 14 13 12 11 10 09 08   2 3 4 5 6

ISBN-13: 978-0-226-47026-9 (cloth)
ISBN-13: 978-0-226-47028-3 (paper)
ISBN-10: 0-226-47026-1 (cloth)
ISBN-10: 0-226-47028-8 (paper)

Library of Congress Cataloging-in-Publication Data

Laszlo, Pierre.
    Citrus : a history / Pierre Laszlo.
        p.   cm.
    Includes bibliographical references and index.
    ISBN-13: 978-0-226-47026-9 (cloth : alk. paper)
    ISBN-10: 0-226-47026-1 (cloth : alk. paper)   1. Citrus.   I. Title.
    SB369 .L37   2007
    634′.304—dc   2007004139

to Muriel and Coleman † Citret, with love

Contents

Acknowledgments

In preparing this book, I have benefited from a network of friendly and cooperative informants (most are academics, Ph.D.'s or professors): Henrique Barreto Nunes, Braga, Portugal; Janine Bénazet, Espeyrac, France; Maria José Boccorny Finatto, Pôrto Alegre, Brazil; Eric Boomhower, Lakeland, Florida; H. de Boubuat, Paris; the late Georges Bram, Orsay, France; Jorge Calado, Lisbon; Muriel Citret, San Francisco; Jennifer Cleland, Ithaca, New York; Carlos Filgueiras, Rio de Janeiro; Betty Fussell, New York; chef Cynthia A. Goldman, Wilmington, Delaware; Raquel Gonçalves Maia, Braga; Monique Gras, Sénergues, France; Chantal Humbert, Souliers, France; Marie-Line Lecourt, Liège, Belgium; Françoise Létoublon, Grenoble; Maria Augusta Lima Cruz, Braga; António Lopes, Lisbon; Fernando Luna, Rio de Janeiro; Hernani Maia, Braga; Diane Middlebrook, Palo Alto, California; Marion Nestle, New York; Eduardo Pires de Oliveira, Braga; the late Pierre Potier, Gif-sur-Yvette, France; Christopher Ritter, Berkeley, California; Roberto Rittner Neto, Campinas, Brazil; Jacques Royer, Paris; Antonio Ruiz Ruiz, Cordoba, Spain; Jody Scott, Southern Pines, North Carolina; Asunção Vasconcelos, Braga.

I am most grateful to the staff of several libraries, especially the Boyd Library at Sandhills Community College in Southern Pines, North Carolina, and the Wellcome Library for the History of Medicine in London, for their help. In this connection, I acknowledge the hospitality and the outstanding facilities for scholars that I enjoyed November 15–December 15, 2001, at the Wellcome Center for the History of Medicine in London. I heartily thank its director,

Harold G. Cook, and Ms. Sally Bragg for this privilege. Some of my research in Southern California was done during a stay at Caltech, for which I am extremely grateful to Professor Sunney I. Chan.

I am extremely grateful to my fellow writer Mary Dearborn for her close reading of the manuscript and for editing it into good English prose. I owe thanks to Norma Roche for her scrupulous and felicitous copyediting. This book was carried by the unflagging enthusiasm of Christie Henry.

I am grateful above all to my wife, Valerie, for all her encouragement and formidable close reading.

Prologue: Letter to a Fellow Writer

Esteemed Colleague:

This is a letter from an admirer. In writing your *Chü Lu* (The Orange Record), you have provided the first monograph in history about citrus trees and citrus fruit. According to your preface, you wrote it starting in the tenth month of the fifth year of the Shun-Hsi period. This was when we, in the West, were still in our so-called Middle Ages. Our knowledge then, as embodied in books, was mostly that of the ancients, and it was more of a philosophical than of a technological nature.

Esteemed Han Yen-Chih, allow me to introduce myself: my name is Pierre Laszlo, and I write in the twelfth month of the sixth year of the twenty-first century. Jacques Chirac is currently president of my country, France. Like you, I am also writing on citrus. Your book, published in our year 1178, sets a splendid example for me to emulate.

In your preface, you stated that your reason for writing your monograph was that, as a man of the north, you were new to the orange-growing areas around Ni-shan. As the new governor of Wênchow in Chekiang, you had the opportunity to become knowledgeable about those fragrant groves. Not only were you a conscientious and more than competent administrator, but forgoing the temptations of politics, you elected to rival and even outshine the scholars. In order to further your people's understanding of agriculture, horticulture, and citriculture under your enlightened supervision, you resolved to start by reading about them.

Yet, even though you were able to line up treatises on plants such as peonies and the litchi, there was no such monograph on citrus. Which is why you decided to write yours. You made it a monument to learning, of the best kind, both theoretical and practical.

———

My experience has been almost the exact opposite of yours. Since the appearance of your book, monographs on citrus have become too numerous to count. Why, then, add to the list?

A few years ago, I published a book on salt. For many centuries, rulers controlled the production and distribution of salt to better tax an ingredient essential to life. They turned this commodity into an instrument of their power and wealth. Likewise, oranges—except in privileged oases such as Ni-shan—remained a privilege of the wealthy and powerful until quite recent times. I was struck by the similarity.

At the time you wrote your outstanding monograph, citrus was no longer exclusively Chinese. Is it because of your book that citrus plants were packed by travelers in their luggage? In any case, citrus seeds and plants began to be exported from Chinese soil, and they have now settled over our entire planet—at least within the climactically favorable belt, with which you are familiar, circling the equator.

First, the whole of Europe became familiar with citrus trees and their cultivation. But their westward expansion did not stop with Europe. The portion of humankind that recorded its deeds in writing then traversed the great ocean to the west of you, and it colonized territories even larger than China, your Empire of the Middle. People discovered that there were also Outer Empires, and they brought citrus plants along as they went west. Citrus became their livelihood. I wanted to chronicle this push to the west, as it affected not only the world's economy, but its cultures as well.

My impetus for writing also differed from yours. You wrote the Orange Record while in active life, and I wrote my book in newly elected retirement, when I could give myself all the time I needed to put together the required documentation—a part of the writing life that I very much enjoy. I also wanted my book to be bathed in an atmosphere of freedom—the freedom I gave myself while writing it, the freedom to follow my whims. Much as you regretted not being able to leave the city and your office for pleasure, during my active life as a scientist I restrained myself from putting pen to paper, except for publishing my scientific results and those of my coworkers.

Moreover, in the words of another Chinese scholar, whose writings we continue to revere, Mencius—whom you probably know as Meng-Tzu:

Me too, I wish to inspire in men honest sentiments, stop the course of bad doctrines, put a brake on self-indulgent acts, ban discourses insulting reason, and thus continue the work of the Three Sages (Iu, Tcheou Koung, and Confucius). Do I like discussion? I cannot abstain from it. Whoever is capable either of refuting or discarding the doctrines of Iang Tchou and of Me Ti is the disciple of these three Great Sages.

———

But please tell me, Mr. Han, how you obtained the information for your highly valuable survey. Did you take notes on your inspection tours? Did you grow citrus on your property to better observe its characteristics and growth patterns? Perhaps you also engaged in the hobby of raising miniature plants in clay pots in your house? And which among these plants was your personal favorite? Should I take it that it remained the *chü* orange from Ni-shan, which started your love affair with the orange blossoms and fruit?

———

Were you to find the time in your busy schedule as governor to send me an answer, I should feel most honored and gratified.

PART ONE

Domestication of Exotic Species

ONE

Introduction, Including a Definition of Citrus Plants and Fruit

Agridulce como la naranja es el sabor de la vida.
(Sour and sweet like the orange is the taste of life.)
SPANISH PROVERB

This book, ostensibly about fruit, has an obvious autobiographical dimension. My own travel in life happens to have followed the westward trail of citrus: I moved from Europe to Brazil and then to the United States.

Fortuitous as it was, determined as it was by economic, technological, and familial contingencies, this journey gave me an education. I was fortunate to be exposed at a young age to such different and endearing cultures. These experiences have left me some lasting lessons.

————

However, while a plethora of self-referential allusions lurk beneath the surface of this book, I have tried not to let the autobiographical gain the upper hand. I have shared the experience of a whole cohort of Europeans born just before the onset of World War II, who experienced the harsh German Occupation and who went on with their lives in its aftermath.

During the war, food was scarce in Grenoble, where my family lived; hence, it was uppermost on our minds. My parents and I barely survived. At one point, about the time of the battle of Stalingrad, the grocers had run out of anything edible to sell. My father planted potatoes in our small plot behind the house, and we rejoiced at the sight of the green shoots and the flowers. We rejoiced too soon. Insects (*doryphores* in French, Colorado beetles in American English) got to the plants before we did. This experience was metaphoric of our collective plight; indeed, *doryphores* was one of the none-too-kind nicknames we gave to the swarming, green-uniformed German soldiers (*Boches*). By August 20, 1944, when American troops liberated Grenoble, we had had nothing but rutabagas to eat for a couple of months. I have retained a durable, lifelong aversion to this vegetable and to its taste, in any form, in any dish.

After the war was over, the food supply in the two countries I had direct experience of, France and England, improved very slowly. In Europe, I had experienced minimal fruit diversity, and the supply was tightly linked to the seasons: cherries in May and June, followed by cantaloupe and peaches during summer, plums and grapes in September, and chestnuts and walnuts in October. In addition, we enjoyed apples and pears throughout the autumn season.

Citrus of any kind? None. Like many a European at that time, I discovered my first oranges, tangerines, and clementines during the holiday season, as Christmas gifts and luxuries hanging from the tree. Traditional Christmas delicacies then also included dried figs and dates from North Africa.

The list of *un*available fruits was long: bananas, pineapples, watermelons, avocados, figs, pomegranates, guavas, mangoes, papayas, and litchis, as well as citrus fruits. I enjoyed my very first grapefruit when I was about six, when, shortly after the liberation of Paris, we visited close family friends there. I recall my awe at the exotic foods on the table in their rue Raynouard apartment.

My enjoyment of citrus (lemons included) as a Christmastime bonanza connected arboriculture and culture. It took me back and forth across the bridge between the material and the mental, external and internal riches, fruit for the mouth and fruit of the imagination.

By the age of ten, I had a keen appreciation of foodstuffs: I knew that food for the body is, also and more importantly, food not only for the mind but also for the soul. I had yet to learn that food habits delineated key cultural differences between Europe and the Americas and, within the New World, between Latin America and North America.

When I arrived in Brazil in July of 1950, I had reached a personal El Dorado. My personal circumstances made me feel like a conquistador.

After the scarcity we had experienced in Europe during the war years, we came upon the Land of Plenty. The New World spelled abundance, and citrus symbolized its natural riches. The cornucopia I experienced upon waking up in a Rio de Janeiro hotel room was not an illusion, it was a fantasy fulfilled. I enjoyed it again and again, visiting a nature lush and wonderful, whether domesticated in the splendid Botanical Garden in Rio or in the wilderness of Serra dos Orgãos, where one need only bend over to collect gems.

But my El Dorado was no Garden of Eden. A no-man's-land replaced well-trodden trails. In Europe I had been familiar with nature: there, one loads a backpack and starts to climb in the direction of one of the handsome mountains of the Alps. In Brazil I found myself looking helplessly at a forbidden outdoors. The basalt is too slippery to climb. The vegetation is overwhelming and bars any attempt at a leisurely stroll. There are dangerous animals, and poisonous snakes lurk everywhere in the wilderness: one's equipment must be adequate to the constant threat of both.

In a later piece of luck, my college education included a four-year course in botany. I learned that lemons and oranges belong to the family Rutaceae. I shall content myself here with stating a few of the characteristics of the genus *Citrus*; I will spare you the botanical technicalities that designate a species as hemianatropous, crassinucellate, or polystemonous. Rutaceae belong to the Rutales, being one among twelve families in that category. They are found throughout tropical areas. There are 150 genera among the Rutaceae, containing about 1,500 species. Citrus plants are mostly shrubs or trees rising three to thirty feet: they are rich in essential oils, and they are perennials. The undivided leaves are nearly always dotted with glands. The bisexual flowers are conspicuous, large, and pure white or pale pink, rich in nectar and fragrantly sweet. Indeed, small insects such as flies and bees readily pollinate them. Their sepals and petals show threefold or fivefold symmetry. And there are two whorls of stamens, the outer opposite the petals. The female part consists of a single compound pistil, made of two to five or more partly or totally fused carpels (a carpel is the vase-shaped female reproductive module with an ovary at its base), leading into a style that connects it with a sticky stigma on the top, which captures grains of pollen. A single plant may bear up to 60,000 flowers, only 1 percent of which will develop into fruit. The fruit, also known as a hesperidium, is roughly spherical in shape, often yellow to russet in color. It has three layers: the skin or peel (also called the rind), the pith, and the pulp. As a rule, the peel can be removed from the two inner layers. The pulp is made of carpels filled with juice sacs (modified hairs). It contains a sweet and/or sour liquid. After blooming, fruit development takes five to eighteen months. Unlike most other types of fruit, citrus fruits can stay on the tree without becoming overripe. Citrus plants—with the possible exception of the desert lime, *Citrus glaucus*, from the Australian bush—are unable to withstand drought.

The diversity of citrus plants and fruits is mind-boggling. There is now hope that determination of their genomes will bring some order to their catalogs. DNA analysis points to a small number of basic species from which all varieties of citrus originate. There are perhaps only three true species: citron, pummelo, and mandarin. Most cultivated varieties are hybrids. Mandarins, also known as tangerines, have given rise to a number of varieties. The orange, grown in China for millennia, is arguably a cross between a mandarin and a pummelo. The grapefruit, which appeared in the Caribbean during the seventeenth century, is in turn a pummelo-orange hybrid. In any case, varieties arise through a combination of spontaneous mutations (leading to so-called sports, sometimes coexisting on a single tree with another variety) and hybridizations (crosses between existing varieties). However, only genetic evidence will provide accurate and precise delineation of the interrelationships between citrus types and varieties.

Globalization is nothing new. Jesuit missionaries were already trailing across the oceans in the sixteenth century. At the time of the Counter-Reformation, other missionaries followed in their footsteps. They too imposed globalization of a sort, to the detriment overall of native cultures and languages. But missionaries were also responsible for back-transfers, including the introduction of the Brazilian Bahia orange into the United States, under the new name of the Washington navel.

In the past, Jesuit-led globalization installed a citrus belt around the world. In present-day globalization, multinational corporations have replaced monastic orders. Business schools are the modern equivalent of seminaries and Jesuit programs. Nowadays, citriculture has become a worldwide industry with the United States as a base. The vast, affluent American market is where corporations launch and develop themselves. From such beginnings, citrus businesses strive for worldwide control and dominion. In their production of citrus, whether fruit or fruit juice, corporations such as Coca-Cola, McDonald's, Sunkist, Florida Fruit, United Fruit, Sucocitrico, and Louis Dreyfus have achieved the highest standards of sterile conditions together with reproducibility of results. In so doing, they have turned nature's diversity into uniformity. They have vastly reduced the appealing variety of tastes and flavors that make tropical fruit so attractive.

———

To anticipate some of the forthcoming pages, I have gathered here some facts and figures:

- A billion citrus trees are currently being cultivated on the planet, one for every six inhabitants.
- At the turn of the twenty-first century, the world production of citrus fruit is close to 100 million tons.
- Global trade in fresh citrus fruit is about 10 million tons per year.
- The life span of a citrus tree is about a century.
- A five-ounce navel orange has about 75 mg of vitamin C.
- Citriculture in Florida is a $9 billion industry.
- Taken together, Florida and Brazil account for 90 percent of the world production of orange juice.
- In the first years of the twentieth century, the advertising tycoon Albert Lasker invented orange juice ("Drink an Orange" was the motto for Sunkist).
- In 1999, Americans drank an average of 5.7 gallons of orange juice per person.
- Natural orange juice currently costs, wholesale, between $5 and $6 per gallon. Hence, the U.S. market alone is about $9 billion.
- During the 1880s, with the citrus-growing bonanza, the population of California increased by 345,000.
- In the year 1930, two thousand distinct brands of California citrus were being commercialized.
- In the United Kingdom, out of twenty-one samples of orange juice examined in 1990, sixteen were adulterated.
- The main adulterants of orange juice are corn syrup and beet sugar.
- Ugli fruit arrived on the North American market in 1942.
- On February 8, 1895, frost killed to the ground almost every orange tree in Florida. Production declined by 97 percent, and sixteen years passed before it recovered to its previous level.
- Grapefruit juice potentiates the action of Viagra.

But let us now turn from numerical data to the history of human migrations and of the attendant dispersion of plants and foodstuffs.

TWO

Transplantation to Europe

What manner o' thing is your crocodile?

. . . .

The elements once out of it, it transmigrates.

SHAKESPEARE, *ANTONY AND CLEOPATRA*

Plants are often disseminated far from their territories of origin. This may happen through natural processes, such as winds, sea currents, or bird droppings carrying seeds, or through chance events, but deliberate human actions disseminate plants as well. For instance, the tulip found in the wild in Greece and Turkey was brought to the Netherlands, where it became an object of speculative frenzy before becoming simply an emblematic domestic flower.

Citrus trees and citrus fruits tell no less interesting stories. They were introduced into Europe after the Indian conquests of Alexander the Great. Their survival and spread in the Mediterranean region was a fringe benefit of the Jewish diaspora.

Plants often owe their location, and sometimes their very existence, to religion. The ginkgo tree is a "living fossil," preserved because it adorned the entrances to Buddhist temples. Mormons, as they fled religious persecution in the eastern United States, brought samples of *Populus tremulus*—the present-day aspens—to the West on their pushcarts and wagons. The planting of the grapevine, *Vinis vitifera*, followed the progress of Christianization in western Europe because wine was required in the Catholic Mass.

Citron trees exist in the wild today in northern India. Citron seeds have been found in Mesopotamian excavations going back to 4000 BC. Citron trees were first cultivated in Persia. The military expeditions of Alexander the Great brought them back to Greece around 300 BC. Greek colonists introduced them into Palestine around 200 BC. A Jewish coin minted in 136 BC shows the fruit on one side. Another image was discovered with Greek and Latin inscriptions in the synagogue of the Jewish colony in Ostia, the harbor of Rome, dating to the first century BC. Because of its role in the autumn Feast of the Tabernacles, the citron followed the Jewish people in their spread around the Mediterranean, with Salerno supplying the fruit to Jewish communities in western and central Europe.

The fruit, mentioned at the beginning of the Christian era by writers such as Theophrastus, Virgil, Dioscorides, and Pliny, was cultivated in Italy in the third century. In the fourth century, barbarian invaders destroyed most of the trees in mainland Italy. They survived on the islands of Corsica, Sardinia, and Sicily.

Spaniards brought the citron to Puerto Rico in 1640 and to St. Augustine, Florida, at an unknown date. As the *etrog* cultivar of Jewish ritual, the citron was first commercially grown in California in 1880. Nowadays, etrog citron trees are grown in the Sacramento Valley. American Jews order their citrons—a single fruit, well-formed and kosher, sells for about $70—from a handful of suppliers in California and Texas.

Citrons are a feature of Jewish religion. They play a symbolic function in the Thanksgiving ritual of the Sukkot—also known as the Feast of the Tabernacles. This seven-day post-harvest festival is held in the fall (during the month of Tishri), five days after Yom Kippur.

The Jewish Feast of the Tabernacles is held in gratitude to the Almighty for the fertility and growth of plants. It also celebrates the survival of the Jews during forty years of wandering in the desert. This ordeal is evoked through the building of a *succah* (plural, *sukkot*), a temporary tent built from branches lashed together. The faithful bring into the succah palm, myrtle, and willow branches bound together, which they wave in praise of God. They pour water into a sacred basin, a symbol for rain for the crops.

Etrog—the Jewish name—is a small-fruited variety of citron, grown especially for the Feast of the Tabernacles. The heart-shaped fruit symbolizes God's benevolence and goodwill toward one's fellow beings. The Sukkot ritual is associated with the colors green (of the palm leaves) and gold (of the citrons).

———

Other citrus plants from Asia probably arrived in Europe by a similar route. I base this conjecture on events during the Age of Justinian. In the year 532 AD, Justinian, the Eastern emperor of Rome, was facing a tough political situation. The old Roman empire in western Europe was beset by barbarians of all kinds: Celts, Vandals, Franks, Goths, Visigoths.

Justinian had a daring vision. He would rebuild the empire, but from the east. He would move its center from Rome to Constantinople—thus named when Constantine was the emperor, in 324–337. Justinian was a general with great military acumen. His armies were able to recapture much of Italy. A series of military victories boded well for Justinian's imperial ambitions.

Disaster struck in the year 542, in the form of bubonic plague. Historians estimate that, within a single year, between 25 and 50 percent of the population of the eastern Roman empire was wiped out. Likewise, a full third of the people in Ethiopia, Egypt, Persia, and Palestine were lost. Arrival of the Black Death in Europe signaled the beginning of the Dark Ages.

Where did the germs come from? From China. They were carried by fleas on rats hidden in the holds of ships. Commercial traffic was intense along the routes between Asia and Europe known as the Silk Road and the Spice Route, which combined overland and overseas itineraries. Trade along these intensely traveled routes served to import a number of plants, as well as pathogens, from Asia to Europe. Plants were imported because they were deemed useful, beautiful, or both. Citrus shrubs and trees were foremost among such imports.

———

If Judaism accounted for the circum-Mediterranean spread of citrons, Islam established a major beachhead for the cultivation of citrus on the Iberian Peninsula.

Citrus fruits, whether citrons or oranges, were known to the Romans, and one might have expected them to spread throughout Europe. In time, citrus trees would have followed Roman legions in their travels and would have been planted in the warmer parts of Europe and North Africa. From there, traders would have taken the fruit to other parts of Europe.

A major diversion from this standard course of events occurred when Arabs invaded and began an occupation of a major chunk of the Iberian Peninsula (present-day Spain and Portugal) that would last several centuries, from the eighth until the end of the fifteenth.

The Muslim beachhead in southern Europe had a huge civilizing influence. The invading Arabs were not only efficient soldiers, but also scholars, scientists, technical wizards, philosophers, poets, and musicians. The Europeans whom they rolled back for hundreds of miles were, with the exception of monks in a few abbeys, illiterate and unrefined. Even a short list of what Arabs gave to European culture during their occupation of Spain and Portugal ought to include transmission of the scholarly Greek texts; the notion of courtly, romantic love; sung poetry, by way of the troubadours in northern Spain and southern France; huge contributions to philosophy and science, such as algebra and alchemy (it is no accident that these names have an Arabic origin); and miscellaneous technologies.

The Arabic territory was known as *al-Andalus*. Its northern borders were heavily fortified in order to protect it from attack by the Christian armies in the north. The Christians and the Muslims did not coexist peacefully or easily. The Christians sought to push the Muslims back across the Mediterranean into North Africa. Conversely, the Muslims sought to enlarge their dominion and occupy more land farther north. At the same time, Muslims and Christians engaged in numerous exchanges, not only of goods but also of cultural practices, across the no-man's-land separating them. War was frequent, but there were also long peaceful periods between times of warfare. And even in wartime, nonstrategic goods and practices were traded across borders.

What the Moors—as the Arabs were known on the peninsula—introduced was an influx of new peoples, such as the Berbers, who came from North Africa; new irrigation techniques, imported from the Middle East, Syria in particular; and new crops, predominantly citrus, vineyards (in spite of the prohibition against wine in the Muslim religion), and wheat.

The Spanish and Portuguese vocabularies were enriched with names for these agricultural imports, providing eloquent evidence of the exchange:

Spanish name	English translation	Arabic origin
zanahoria	carrot	isfannâriya
chirivia	parsnip	jiriwiyya
azufaifa	jujube	al-zufayzaf
arroz	rice	al-ruz
algarrobo	carob	al-kharrûba
albaricoque	apricot	al-barqûq
		(Continued)

15

(Continued)

Spanish name	English translation	Arabic origin
azafran	saffron	al-za'farân
azúcar	sugar	al-sukkar
berenjena	eggplant	bâdhinjâna
aceituna	olive	al-zaitûna
alcachofa	artichoke	al-karshuf
limón	lemon	laimûn
naranja	orange	nâranjâ
toronia	grapefruit	turunja

Irrigation was mandatory in the semiarid climate of southern Spain. As in Judaism, orange groves were planted for religious reasons. Their economic success had social and geographic consequences. The Koran repeatedly stresses that Paradise is gardenlike. The Andalusian gardens had a model plan: they were divided into four rectangular subplots with rows of trees parallel to an axial water-supplying canal, a design that was imported directly from Persia.

Because the settlers of al-Andalus came from an irrigated land, it was second nature for them to use familiar procedures for irrigation. The Egyptians who settled in Murcia, for example, applied what they knew from the Nile Valley to the basin of the Segura River.

The settlers in the region of Valencia, on the other hand, were Berbers from North Africa, who were not used to tapping water from large rivers. They seem to have been taught how to irrigate agricultural land during the first quarter of the ninth century by an Umayyad governor from Syria. In the system of irrigation imported into North Africa from the Ghûta (the irrigated water belt around Damascus), a river was split into branches, which were then subdivided into a number of canals. The irrigation scheme was set up so that the division of the canals followed the division of time, that of the day into twenty-four hours. When water was plentiful, each canal drew water from the river according to its capacity; but in times of drought, gates were opened and closed so that the individual canals would draw water in succession. In the same way, irrigators were allowed to open their gates as they pleased when the canal was full. Otherwise, they had to wait their turn. And each irrigator was responsible for the upkeep of the portion of the canal that abutted his property.

The latter part of the Andalusian period, from the thirteenth century onward, brought major improvements to this irrigation system. In the *noria* system, a pack animal turned a wheel that lifted a cable with buckets attached. In other places, water-powered wheels, instead of the simple flow of gravity, allowed water into the irrigation canals.

Irrigation and the attendant agricultural prosperity enriched a particular social class in al-Andalus, the urban merchants. They invested their profits in land in the countryside around the cities. This led gradually to the dominance of towns over the countryside, and to the landscape feature known as the *huerta*; that is, a town surrounded by a belt of irrigated fields.

Thus, ten centuries ago, Spain experienced a thriving economy fueled by irrigated agriculture and arboriculture. Irrigated cultures led to the flourishing of numerous cities (Seville, Valencia, Granada, Zaragoza), each with its own citrus belt. In the twentieth century, Southern California had a similar organization, first in the citrus belt between Pasadena and Riverside, leading to the growth of greater Los Angeles, and later on in the Central Valley, between the Sacramento and San Joaquin rivers.

————

The military clashes between Muslims and Christians on the Iberian Peninsula are now history. A hybrid culture was their result—including citriculture.

The arrival of citrus trees in Spain and Portugal was a fait accompli by the eleventh or twelfth century. The great mosque in Cordoba had a spectacular courtyard planted with orange trees, the Patio of the Oranges. In Seville, the mosque has a Court of the Oranges.

Even after the Moors were finally ejected from Portugal and Spain, Christians would not readily identify themselves with the exotic citrus trees, which they still associated with the Muslims. They had to purify them first, not wanting to incur blame for borrowing from their worst enemy.

The Moors never occupied the north of what would become the kingdom of Portugal in 1128, when the Arabic presence on the peninsula had started to wilt militarily and shrink geographically. It was not until the reign of Isabella and Ferdinand in Spain, during the fifteenth century, however, that the Arabs were finally expelled. Indeed, there was never any Arabic presence in the northwestern area of the peninsula between Porto and Santiago de Compostela, in the (present) Portuguese province of Minho and in the adjacent Spanish province of Galicia.

Citrus trees appear to have been introduced into northern Portugal rather late, toward the end of the fifteenth and the beginning of the sixteenth century. For instance, João de Coimbra planted sixty orange trees when he renovated an aristocratic house in Braga in 1512. He was a powerful person, as *provisor* and vicar-general of the archbishopric there. In the same Portuguese province of Minho, the legendary first orange trees, planted earlier, are credited to similarly respected and even saintly religious officials. One such story ascribes the importation of orange trees into Portugal to the Benedictine Cistercian monks from the monastery of Bouro (which lasted from 1148 to 1834). The tale has a factual basis. A nineteenth-century description of the convent mentions its sweet oranges, which must have been planted after 1567, the date for the architectural renovation of its buildings.

Also according to tradition, Dom Frei Bartolomeu dos Mártires (1514–1590), who became a Dominican in 1528 and was elevated to archbishop of Braga in 1558, planted an orange tree—presumably during his tenure as archbishop—at the Tower of Lanhelas, a village between Caminha and Vila Nova de Cerveira. He was a highly respected figure, canonized by Pope John Paul II.

———

There are many other examples of the "Christianization" of citrus trees and citrus fruit throughout the Iberian Peninsula and Italy. During the Middle Ages, lemons and oranges spread over all of southern Europe. Because orange trees come to fruition during winter, they became part of rituals celebrating renewal and the hoped-for return of spring in pagan times; later, they were involved in Christmas festivities. Oranges at Christmas were available only to the affluent and powerful, becoming common only as conditions improved with the Industrial Revolution. Oranges and tangerines served as decoration, doubling as precious delicacies coveted as gifts by children.

One of my memories as a child in Grenoble is being allowed to stay up late after the end of the Christmas dinner and being encouraged by my parents to remove the citrus fruits hanging on the tree to indulge in them. My siblings and I then played with the peels. It was fun to squirt the tiny droplets into the flame of a candle. One would glimpse a tiny explosion as the volatile oils would ignite and burst into flame. (This may have been one of the earliest signs of my future career as a chemist.)

Indeed, in France in the 1940s and early 1950s, one could purchase citrus from a greengrocer only during the holiday season, which lasted only a few weeks, or at most until March or April. Hanging oranges from southern Europe in the Christmas tree turned it into another, more exotic evergreen—an orange tree, as it were. Each household, whatever its means, thus gave itself the illusion of princely ownership of fruit-bearing orange trees during the cold and otherwise sterile season. Only during the 1950s, and to a greater extent, during the aptly named Golden Sixties did the golden fruits start becoming commercially available in provincial France year-round as supermarkets replaced local markets.

———

Besides religion, the other great force in cultural history, also with lasting influence, is economics. Citrus fruits and trees became valuable in trade very early on. As food, the fruits were coveted delicacies. Their peels would also become an object of commerce, as medieval French or English cookbooks such as the *Mesnagier de Paris* attest. The heavenly smell of orange flowers and the gorgeous perfume of ripe oranges or tangerines made them valued luxuries.

The bergamot orange is a case in point. Its trajectory over the centuries includes regions as distant from one another as Calabria and the Ivory Coast. The spread of the bergamot boasts legendary components, some of them obviously erroneous. The bergamot is a bitter orange, whose aromatic peel is used in both cologne and Earl Grey tea.

The word "bergamot" itself is already something of a puzzle. According to legend, Christopher Columbus brought the plant back from the West Indies or the Canary Islands to Spain. It then reached Calabria from the town of Berga, near Barcelona, where it took the name *bergamot. Si non é vero, bene trovato.* (If this is not true, it is well imagined.) In fact, the word stems from the Italian *bergamotta*, no relation to the Italian city of Bergamo (nor to Berga). It derives instead from the Turkish *beg-armûdì*, with the literal meaning of "the lord's pear"—which points to a Turkish or Central Asian origin. The French name *bergamotte* was first used in 1536 for a variety of pear. It came to apply to a citrus variety around 1699.

By that time, the essential oil of bergamot had become a key ingredient of cologne. Its discoverer, Giovanni Paolo Feminis, was a Piedmontese peddler, born in Crana in 1666. In the early 1680s, he plied his trade in the Rhine Valley, near Cologne—the Roman *Colonia*, German, Köln. Among his wares, he pushed an *acqua mirabilis*, a "wonder water" made of distilled wine and the essential oils of bergamot, lavender, and rosemary. Bergamot oil gave the mixture a top note of freshness while fixing the aromatic bouquet, enhancing and softening the other contributions.

Feminis claimed to have acquired the secret recipe for his cologne water from a monk. He was peddling it as a cure-all, useful for stomachache and bleeding gums in particular. *Acqua mirabilis* made his fortune.

A relative of his, Giovanni Maria Farina, born in Santa Maria Maggiore in 1685—Feminis's mother was the cousin of one of Farina's grandmothers—joined him in Cologne. Feminis died in 1736: by 1732, Farina had taken over the marketing of Cologne water. This was the time of the Seven Years' War in central Europe, and soldiers going home brought back Cologne water. Comtesse du Barry (1741–1793), one of the numerous mistresses of the French king Louis XV, brought it into fashion. Later on, Napoleon splashed himself with it, or bathed in water containing a flask of it.

So-called secret recipes are often ascribed to monks, pirates, or wise Asians. The latter is the case with Earl Grey tea: the earl is said to have been given the recipe by a Chinese mandarin with whom he was friends. Charles Grey (1764–1845) was prime minister of England in the early nineteenth century, at the time of the Napoleonic Wars with France. Originally made from unsteamed China black tea and flavored with oil of bergamot, Earl Grey is now made from India and Ceylon (Sri Lanka) black teas together with oil of bergamot. Cologne and Earl Grey tea made bergamot oranges a profitable harvest from the mid-eighteenth century on.

During the whole period between 1750 and 1950, bergamot trees were grown almost entirely in Calabria, in a narrow belt with a total area of about 1,300 hectares, along the Tyrrhenian Sea from Punta Pezzo to Capo d'Armi and along the Ionian Sea from Capo d'Armi to Punta Stilo. Small farms, one hectare on average, each brought in a tiny revenue of about seventeen thousand present-day dollars yearly.

As early as 1815, there were attempts to acclimatize the plant in the United States: in Louisiana, Florida, and California. They all failed. Attempts to acclimatize the plant in other countries—in South America, North Africa, Sicily, and French Guinea—also failed, as these places failed to provide the right climate. Bergamots are demanding, and seem to be happy only in Calabria.

However, cultivation of bergamots in Calabria fell victim to the same factors that would drive orange growers out of the Pasadena-Riverside citrus belt in Southern California and into the Central Valley: urbanization, industrialization, the rise in real estate values, and tourism extending into the whole sunny Reggio Calabria area.

The global demand for bergamot oil is on the order of about 120 tons per year, which translates into 24,000 tons of fresh fruit (the oil amounts to 0.5 percent of the fresh fruit). Since bergamot oil may sell wholesale for about $170 per kilogram during good years, it can become a profitable crop. Given this potential, in the mid-twentieth century, a few French planters started an agribusiness in the Ivory Coast (Côte d'Ivoire, a West African country) by planting bergamot in the region of Sassandra. At last, another favorable climate for the bergamot's growth had been located. The Ivory Coast government further developed cultivation in 1970 by financing a *Consortium des Agrumes de Côte d'Ivoire*. The plantations there are much bigger than in Italy, averaging 50–250 hectares as a rule. Their yearly revenue is thus in the hundreds of thousands of dollars. The key to high profits—as with any chemical product, whether commodity or fine chemical, natural product or synthetic—is a match between capacity and demand. If the price of bergamot oil shoots up, then the buyers (predominantly in the areas of perfumes, cosmetics, and foods) go for cheaper substitutes. They replace the natural oil with its key ingredients, linalool and linalyl acetate, which are synthetic and considerably less expensive.

––––––

Bergamot oranges continue to be grown in Calabria. If one goes across the Strait of Messina into Sicily, one encounters another orange variety. This is a sweet, not a bitter, fruit. It is attractive not only to the mouth and the nose, but also to the eye, and to the inward eye of imagination and fantasy: its juice is blood red.

Sicily is a triangular island facing the Calabrian toe of the Italian boot across the Strait of Messina. Over the centuries, invaders came from every direction of the compass and occupied the island. Hence, Sicily is culturally rich. There are Greek temples, such as those in Agrigento. Vikings came from the North. Arabs came from the South. Both of these waves also left monuments, such as the gorgeous Monreale Palace in Palermo. Immigrants also came from areas around the Mediterranean—Albanians, for instance.

Whereas an island such as Sicily can become a melting pot of cultures, its territory is naturally isolated. One recalls Charles Darwin's observations of singular varieties of several animal species—finches, tortoises, iguanas—on individual islands in the Galápagos Archipelago. Insularity may account likewise for the appearance of blood oranges in Sicily as an accidental variety—as a "sport," to use the terminology of biologists. When did they first appear there? It would seem to be several centuries ago, but exactly when remains in doubt.

The product of this mutation has turned into a significant economic resource. Blood oranges account for about 60 percent of Italian citrus production and fetch a good price on European markets. Consumers consider them a delicacy. During the long, dreary, gray, often rainy and cold European winter, the three main varieties (Sanguinello, Moro, and Tarocco) bring their sunniness to any meal in which they serve as a dessert treat.

Like the setting sun, the fruit is reddish orange. Its outer color is an excellent predictor of its inner hues. These oranges are small or medium in size, and they carry seeds and a blood-colored juice that usually tastes a little more tart than that of ordinary oranges—Hamlin or Valencia oranges, for example.

Blood oranges owe their color to dye molecules, often improperly referred to as pigments. A pigment is an insoluble solid, in contrast to a dye, which is soluble and colors the liquid dissolving it. These blood-colored molecules belong to the same chemical families that are responsible for the color of, for instance, cherries, strawberries, and raspberries. Some of these dyes are carotenoids, and related molecules account for the bright color of foliage in northeastern North America during autumn. As the fruit matures, the amount of dye increases about fivefold, which accounts for the development of color.

Indeed, the same climatic factors that contribute to Jack Frost's autumn display explain the presence of these dyes. Citrus fruit in a tropical or subtropical climate devoid of sharp winter-summer alternation tends to remain lime green—chlorophyll green, actually. Citrus fruit growing in a temperate climate with sharply defined seasons first turns orange and tends to redden upon further maturation. This genetic trait has accidentally become selected in the blood oranges of Sicily. On the terraced orchards on the slopes of Mt. Etna in eastern Sicily where blood oranges are grown, cold winter nights alternating with mild days are the cue for the fruit to blush. Besides carotenoids, brightly colored anthocyanins, some red, some purple, and some violet, give these fruits their berrylike taste and color.

They are not only a joy to the eye and a pleasure in the mouth, they are good for the health too. The evidence is widespread. As an example, a comparative study was done in Toulouse, close to where I live, and in Belfast, Northern Ireland. The Toulouse dwellers had a much lower incidence of coronary heart disease, which was correlated with the presence in their blood of twice as many hydroxy carotenoids, from citrus fruits in particular. Blood orange juice has antioxidant activity associated with the anthocyanin dye molecules. These molecules have been shown to protect against certain types of cancer as well.

The name "blood oranges" emphasizes kinship with human blood. Cannibalism? One might experience a slight sense of transgression when partaking of these relatively rare fruits. There is a sexual undertone too, which a minor French poet, Jacques Prévert, thus expressed:

Blood orange
pretty fruit
tit of your breast
drew a new line of fortune
in my hand.

———

Blood oranges arose from a spontaneous mutation. Another citrus fruit, likewise a mutation but more recent, is the clementine. It was also found in the Mediterranean cradle, not in Spain or in Italy, but rather across the sea in North Africa. Since this chapter opened with the role of citrons in Jewish religion, it may be appropriate to close it with clementines, discovered by a Christian monk—though in this case the ascription is not fanciful.

Clementines are named for Father Clément Rodier, who invented the fruit. That is the conventional wisdom, as imparted by most sources. Lucky Rodier, who left his name to a fruit. Other fruits in the citrus family have been named for a language (mandarin), a country (Portugal was the name of sweet oranges around the Mediterranean for centuries), and various cities (Seville, Tanger, Washington, DC).

As is often the case, the truth is more subtle. Rodier did not give his first name to this fruit: he had given up his first name when he made his vows and became a monk. Clément was neither his first name nor his family name. It became his given name when he entered a religious order.

What was his first name, then? Where was he born? In which year? Secondary sources do not answer these questions satisfactorily. The food historian has to do a bit of work in order to sort such questions out and get to the truth about Rodier.

He was born Vital Rodier, on May 25, 1839, in Chambon, in the canton of St. Germain l'Herm, in the *département* of Puy-de-Dôme (in south central France). He became a member of the Order of Saint-Esprit and was sent to Algeria to evangelize the Muslims. He and his fellow monks ran an orphanage in Misserghin, near Oran. The date of the discovery of the fruit bearing his name is also lost in the historical haze. We can only surmise that it happened between 1892 and 1902. The latter date is when the name *clémentine* was approved as an *appellation contrôlée*, in like manner to the best French wines, by the Agricultural Society in Algiers, which also awarded a gold medal, probably handed out to the monk heading Rodier's congregation.

Similarly, we know neither the circumstances of the discovery nor the parentage of the new fruit. And we do not know the part Rodier played in this discovery. Did he come upon a natural hybrid? Was he responsible for the cross? The experts believe that clementines are hybrids between an ordinary tangerine, of the variety *Citrus deliciosa* Tenore, and an ornamental variety of bitter orange, *Citrus aurantium* L., the "Granito."

―――――

The example of the clementine shows that the task of writing history is fraught with difficulties. Even though barely a century has elapsed since the fruit's discovery, there are no written testimonies, and any witnesses are dead. Moreover, the Fathers of Saint-Esprit do not open their archives except to ecclesiastical historians. Most of the evidence has become hearsay. What, then, can we surmise?

In this particular case, I will confess to bias. I am influenced by the personality of the former village priest in the French village where I live. Trees were his hobby. He loved to raise them, he loved to prune them, and he loved to graft them. You could have mistaken him in appearance for any of the farmers in his area. His was a simple soul: he was an intelligent person, but no intellectual. He read little outside of the Bible, his missal, and the newspaper.

I see Vital Rodier, otherwise known as Brother Clément, as a similarly simple, endearing soul. He was, perhaps, given the responsibility of looking after the vegetable garden for his small religious community. There may have been a citrus grove that included various types of trees. Brother Clément must have loved working in the garden, as it was easier to deal with than orphaned children, who can become unruly at times. A contemporary of Gregor Mendel, also a monk, did he also experiment with plants, citrus trees rather than peas? Such an inclination would tie in nicely with the name *clémentine*, honoring him.

Today, most clementines come from Morocco, followed by Tunisia, Algeria, and Spain. The fruit is grown almost exclusively in Morocco, though it was introduced into the United States in 1909 and brought to California in 1914.

―――――

This chapter has focused on three citrus varieties: citrons, bergamot oranges, and clementines. All three were brought to Europe by Semites—Jews and Arabs. The Jewish religion was responsible for the spread of the citron. Arab traders brought lemon and orange trees from the Far East. Arabs subsequently colonized and civilized the southern Iberian Peninsula and northern Africa, and they introduced irrigation, agriculture, and citriculture to those lands. The culture of bergamots in southern Italy and blood oranges in Sicily was, similarly, probably derived from Arabic beginnings.

A robust conclusion that can be drawn from this chapter is the genomic plasticity of citrus. Whether their easy hybridization is a spontaneous feature existing in the wild or a man-made development, the sheer number of resulting sports—in the thousands—is mind-boggling. We owe fruits as tasty, fragrant, and appetizing as lemons, clementines, bergamots, and blood oranges to such profuse vitality. The metaphor of the cornucopia is inescapable. Greek mythology had another, related metaphor, the garden of the Hesperides, with its golden apples. Citrus fruits are bountiful. They offer themselves in endless variety.

THREE

Acclimatization to the New World

El amor y la naranja / se parecen un poquito / por más dulce que ésta sea / siempre tiene su agriecito.
(Love and the orange / resemble one another / however sweet / it always remains a little sour too.)
FOLK SONG FROM ARGENTINA

At the beginning of the twenty-first century, red wines from Argentina and Chile, made from Malbec grapes, enjoyed worldwide popularity. To the dismay of the French wine industry, their success has threatened the primacy of French wines. Ironically, this particular grape variety originated at the end of the nineteenth century in the area near Cahors, known as Quercy. French immigrants transplanted it to South America.

In the same way that red wines from Argentina and Chile now rival the red wines of France, Italy, and Spain, citrus fruits grown in the New World—ultimately in Brazil, Florida, and California—have gradually come to dominate worldwide cultivation. These areas' production of oranges, grapefruits, lemons, limes, and many other varieties of citrus outstrips that of the rest of the world.

Such stories are commonplace. Travelers bring plants for cultivation to new lands far from their area of origin. As Fernand Braudel pointed out in a seminal article in 1961, such transfers are the stuff cultural exchanges between civilizations are made of. They root economic history in the history of travel and discovery. As we shall see in this chapter, if California and Florida are giants in present-day citriculture, it is because the first Spanish settlers planted citrus trees there soon after their arrival. Indeed, ever since the discovery of the New World in the sixteenth century, Spanish and Portuguese conquistadors brought plants to be turned into remunerative crops there (sugarcane, citrus trees, and later on, others, such as cotton and coffee). They cultivated native cocoa as well and transported slave labor for these plantations.

Such is the standard account. It is both what one expects and part of the conventional wisdom. The actual story is considerably more interesting. The added surprise is that some of the main episodes are relatively recent, and that certain individuals have played a key role. The introduction of the potato into France by Parmentier is a good analogy to the citrus tale. If parts of the story are mythical, others are historical, whereas one might have expected the whole story of citrus transplantation from the Old to the New World to be anonymous and mythical.

———

Brazil is one of the major players on the citrus world stage. Brazil alone produces more than 50 percent of the total orange juice concentrate in the world. History, in this case, explains the economics. From the very beginning of the sixteenth century—the Portuguese navigator and explorer Pedro Alvares Cabral discovered Brazil in 1500—the Portuguese planted citrus trees on the coast of Brazil.

The Portuguese, at the beginning of the sixteenth century, used their overseas territories for banishment. A person who had displeased the king might be deported to a faraway part of the worldwide Portuguese empire. He was known as a *degradado*; that is, "degraded," having fallen from royal favor.

Thus it was with Fernando Lopes, a castaway on St. Helena, in the South Atlantic Ocean (later to be made famous or infamous by Napoleon, also in exile). Dropped there in 1513, on a voyage back from India, Fernando Lopes became the first known inhabitant of the island. Alfonso d'Albuquerque (1453–1515), Portuguese viceroy of the Indies, had ordered that Lopes, as a traitor to his country, be mutilated. Rather than returning to Portugal in his humiliating maimed condition, Lopes opted for being marooned on St. Helena with three or four slaves. A few years afterward, the king of Portugal ordered Lopes brought home, presumably to see evidence of the mutilation. He then sent Lopes back to St. Helena, where he remained stranded until his death in 1546.

Lopes had fellow sufferers who were banished to Brazil, and the journey of citrus to the New World is intertwined with theirs. That story has a nameless hero; the identity of this mysterious man remains unknown. The reasons for his banishment pose a historical puzzle. He was a well-educated person, perhaps even an intellectual or a scholar. He seems to have been an astute developer, and he planted many citrus trees. He seems to have become wealthy in the process. The story also hints that he had a ravenous sexual appetite, for he had numerous children with local Indian women. (The Portuguese were well known for their policy of mixing with indigenous women so as to populate their newly acquired colonial territories quickly.) The story takes place on the coast of the state of São Paulo, now the most densely populated and the richest region of Brazil, which to this day continues to provide the bulk of the Brazilian citrus crop.

The bare outlines of the story are known. In 1501, just after the discovery of Brazil the year before, a fleet under Gonçalvo Coelho explored some 3,200 km (2,000 miles) of the Brazilian coast. Among the crew on these ships, one man would later become famous: the Florentine cartographer Amerigo Vespucci, for whom the new continent would be named. The boats also carried a group of men whom, for various reasons, the king of Portugal had deported. One of those convicts was dropped off on the island of Cananéia.

The castaway is remembered in history as *O bacharel de Cananéia*, which translates as "The Cananéia Graduate" or "The Canaan Sage"; in other words, an educated man, since the term *bacharel* refers to a degree similar to the modern Bachelor of Arts or Sciences. It is highly likely that this educated person was a Jew, and that this was the reason for his expulsion from Portugal. Had he refused to renounce his religion? Was he one of those Jews expelled from Spain in 1492, who had found at least temporary refuge in Portugal? Was he one of the New Christians—that is, someone forced to convert and who, after having been deprived of his position, had to leave the country?

Most historians identify the Canaan Sage as one Cosme Fernandes. He was indeed a *degradado*, cast away during the 1502 voyage of Gaspar de Lemos and Amerigo Vespucci. Later on, both Spanish and Portuguese navigators would drop anchor at São Vicente, not far from Cananéia, where Fernandes was living and building brigs. Cosme Fernandes became a wealthy and powerful person. For instance, in a single deal, he sold no fewer than six hundred Indian slaves to a Spanish navigator, Diogo Garcia de Monguer.

The captain in administrative charge of São Vicente, Martim Alfonso, had a tug of war with Fernandes, who failed to heed the banishment order confining him to Cananéia. The royal decree was thus renewed in 1531, and the Sage was forced to depart São Vicente, with his huge family, in July of 1531. The reason may have been that the Portuguese sovereign did not like to have a Jew (perhaps a Freemason as well) enjoying authority over one of his possessions.

Some authors claim that the expulsion of Fernandes from São Vicente was due to an intrigue in the royal palace. Another Jew, Henrique Montes, a veteran of an expedition to Brazil led by a Spaniard, Juan Dias de Sólis, coveted the wealth of Fernandes. Hence, Montes hinted to the king, Dom João III, that Fernandes's wealth was illegal, since he had ignored the king's orders to remain in Cananéia. The king recalled one of Fernandes's sons-in-law, Gonçalo da Costa, to Lisbon, entrusting him with the leadership of a flotilla that sailed to São Vicente and removed Fernandes from power.

––––––

The Canaan Sage may have been a precursor of another wave of migrants. When, at the end of the sixteenth century, the Inquisition in Portugal resumed with renewed vigor and hounded Jews and recent converts alike, some of those hunted people sought shelter back in Spain. But others emigrated to the New World—sometimes under a new identity, some even trying to change their looks. Many went to Brazil. During the first few decades of the eighteenth century, most of the European population in Brazil had such an origin. These migrants formed an urban community in places such as Rio de Janeiro, which included many lawyers and physicians. An unanswered question is whether this educated segment of the population became Christian converts, or if they continued to practice their Judaism in utter secrecy.

The mysterious Canaan Sage is the likely father of Brazilian citriculture. Travelers to Brazil in subsequent years would comment on the abundance of citrus trees there: Rodolfo Garcia, a Spanish visitor in 1540, made a note of the oranges, lemons, and citrons being grown. Jean de Léry, who came to Brazil in 1555, commented that the oranges and lemons that the Portuguese had planted were sweet and huge, the size of at least two fists put together.

The Canaan Sage story, obscured as it is by historical haze, has become one of the founding myths of Brazil: the European castaway; the seducer of local Indian women, fathering dozens of children; the planter, starting a colonial economy of arboriculture. All of these are more or less stereotypes of Brazilian manhood, of the world of machismo that is still very much alive today. In addition, the myth singles out the Sage, *O Bacharel de Cananéia*, as a learned, scholarly person, in contrast to the ruffians and military adventurers who were his contemporaries. The phrase *Bacharel de Cananéia* is formulaic; it abstracts from history mostly forgotten events and persons and creates from them a mythical and heroic figure.

Brazilian plantations—citrus plantations included—thrived on the availability of cheap labor in the form of slaves. The earliest known shipment of African slaves arrived in Brazil in 1538.

The figures are impressive: during the period 1451–1600, 49,500 African slaves, 18 percent of those brought to the Americas, went to Brazil. During subsequent periods, these numbers increased greatly: during the period 1601–1700, 41 percent (549,810) of the total slave population were taken to Brazil; during the period 1701–1810, 31 percent (1,876,120) were sent to Brazil; and finally, during the period 1811–1870, 60 percent of the total of enslaved Africans (1,138,800) arrived in Brazil.

Brazilian slavery differed not in kind, but in degree, from forced labor in the Caribbean or in the British colonies of the American South. Slavery, to the British capitalists, meant first and foremost economic profit from the low-cost labor underwriting sugar and cotton shipped from the American colonies. The West Indies were especially profitable. Slavery in these regions was probably both harsher and more mindful of the well-being of the slaves because of the stronger profit motive. Miscegenation was rare.

For the Portuguese colonists in Brazil—in truth, no less ruthless to their slaves—slavery was primarily run on patriarchal lines, reproducing the organization of the country itself. Each plantation owner was a god to everyone in his household. Miscegenation was common, if not the rule. It was a means for a mixed Afro-Mediterranean society to thrive in the tropics.

African slaves were also transported to the West Indies in very large numbers. The Portuguese (as the enduring Papiamento language on some of the Caribbean islands testifies), the French, and the British were active in this trade. Citriculture had an early start in the Caribbean as well. Christopher Columbus had brought lime seeds on his 1493 voyage. Citrus was outstripped by sugarcane as a profitable investment a couple of centuries later. But, in addition to its impact on the ecology of the islands, there are quite a few cultural traces of citriculture in the West Indies. For instance, in 1671, Jean Estiemble lived in an area of Grande Terre, on the French island of Guadeloupe, called the Citronniers—that is, the Lemon Trees.

Given the fact that citrus trees and citriculture were introduced so long ago to the West Indies, it is difficult to document which variety was first planted where. In some cases the story has become quasi-mythical, to the point, sometimes, of farce.

The story of the grapefruit—also known as the shaddock—is one such example. We know, for instance, that a certain Captain Shaddock, an English sea captain with the East India Company, brought seeds of this fruit with him to the West Indies in 1683 from the Malay Archipelago. A contemporary, Hans Sloane, wrote in his 1707 *A Voyage to the Islands Madera, Barbados, Nieves, S. Christophers and Jamaica*, "In Barbados the shaddocks surpass those of Jamaica in goodness. The seed of this was first brought to Barbados by one Captain Shaddock, Commander of an East India ship, who touch'd at that Island in his Passage to England, and left the Seed there." From a November 9, 1683, issue of *The Kingston Times and Herald*, we learn that the captain hand-delivered some seeds from an Asian plant or tree to William Jones, a planter from the Mandeville area.

But that is about all we know. We can only speculate whether Captain Shaddock might have been the brother of John Shattuck, a grandson of Bermuda's governor who lived in Essex County, near Coggeshall. The spelling is different, of course, and even then we are on shaky ground, for, as Joseph Hunter wrote in his 1843 *The Founders of New Plymouth*, "The mere possession of a surname which coincides with that of an English family is no proof of connection to that family."

And where did the grapefruit get its several names? It's called a *pommelo*—or is it *pomelo*, or *pummlo*? It is also called the pamplemousse, Bali lemon, Limau besar, and shaddock. The source of the last name is obvious, but even then, questions remain. Was the British East India Company, as seems likely, responsible for Shaddock's journey, or was it its Dutch namesake and, later, competitor? For that matter, was Shaddock on a mission from the Crown to explore the Malay Archipelago for possible crops to bring to the Caribbean, or was it his own idea? How long did it take for cultivation of this fruit to spread around the Caribbean and, later, to Florida?

The story one can imagine about our captain is evocative indeed. We can picture him with a ruddy face, perhaps scarred by a saber cut and softened by bushy white whiskers, presenting himself dutifully to his host, perhaps bearing a gift of antique Chinese armor and a tea set of Chinese porcelain for his wife, along with a pair of ivory-inlaid, sterling silver spoons. We can imagine the tales he might have told about his voyage, which might (or might not) have involved great risks and escapes from calamity.

Or we can imagine Captain Shaddock as a corporate officer for the East India Company, telling investors that owning stock in the grapefruit as part of a diversified portfolio would be a wise choice. Captain Shaddock might (or might not) have been greeted in the West Indies as someone bearing a fruit that seemed to be a most promising capital-intensive venture. Whether the grapefruit would turn out to be a worthwhile investment or just a lure—comparable to lapis lazuli in the mines of Trebizond, or vanilla beans on the islands of the Azores—treasures pursued by so many investors that the event is still known as the "Black Gold Rush," as a Liverpool broadsheet of 1683 reported—was part of the risk planters (or "investors") took in planting Shaddock's seeds.

But Captain Shaddock did exist, and it's possible to say that the grapefruit first arose in the eighteenth century on some Caribbean island—maybe Jamaica, but maybe not—as a spontaneous hybrid of a sweet orange and a pummelo (the latter was probably the species brought to the West Indies by the captain). It's called a grapefruit because of the grapelike clusters of fruit on the tree. Its first recorded mention was in 1750 by Griffith Hughes, rector of St. Lucy's parish in Barbados, where it was known as the "forbidden fruit"—even though Madison Avenue advertising had yet to be invented.

The grapefruit was introduced into Florida by Count Philippe Odet, a Frenchman who settled near Safety Harbor on Tampa Bay in 1823, bringing with him the seeds or seedlings of grapefruit and other citrus fruits from the Bahamas. Cultivation began in a number of American locations, in Texas as well as in Florida.

The first seedless variety of grapefruit, Marsh, was fortuitously discovered around 1860 on a farm near Lakeland, Florida. Next came the Thompson variant, a limb sport on a Marsh tree in an orchard owned by W. B. Thompson in Oneco, Florida. Discovered in 1913 by S. A. Collins, the Thompson was introduced in 1924 by the Royal Palms Nurseries of Oneco. The Thompson was the first pigmented grapefruit, with flesh and rind reddish orange rather than citrine in color. Then the Texans moved in with the Redblush, a limb sport of the Thompson, first observed in 1931 by J. B. Webb of Donna, Texas, and introduced in 1934. However, the Redblush was not quite the ultimate grapefruit, as we shall see at the end of this chapter.

––––––

Sometimes the business acumen of a utopian is peerless, as demonstrated by the large-scale planting of limes in the Caribbean. Sour limes (the kind most often found in the grocery store) originated in northeastern India and were brought by the Arabs to al-Andalus. Sour lime trees were planted in Italy around the twelfth or thirteenth century, presumably by returning crusaders. Spanish and Portuguese navigators in the sixteenth century brought the sour lime to the Americas, where plantations were started and where some of the plants escaped into the wilderness in places such as southern Florida and the Caribbean.

Caribbean lime plantations started supplying the British market during the mid-nineteenth century. Joseph Sturge (1793–1859) was an English entrepreneur and philanthropist who held radical political convictions for his time. A Quaker, he was thus a pacifist. He even traveled to St. Petersburg in a vain attempt to prevent the onset of the Crimean War. He was also a strong believer in universal suffrage. He not only campaigned against slavery, but also made sure that his own practices were consistent with his beliefs.

Around 1857, Sturge bought the Elberton Sugar Estate on the Caribbean island of Montserrat. His family business in Birmingham imported lime juice from Sicily for the manufacture and sale of citric acid. But the citrus crops in Sicily failed, and an alternative source had to be found. In 1860, after Joseph's death, his son, John Edmund Sturge (1842–1880), then eighteen years old, and his nephew, J. Marshall Sturge, went to Montserrat. There they sought the advice of Francis Burke, who already owned a lime tree plantation on the island. Burke helped the Sturge family to replace their sugarcane with limes, making sure that only free labor would be used in the process.

––––––

To most Americans, the key lime name evokes key lime pie. Key lime trees (*Citrus aurantifolia* Swingle) are extremely thorny shrubs. The fruit is among the smallest in the citrus family, about the size of a ping-pong ball. The peel is thin and smooth, and is yellow green at maturity. The fruit contains ten to twelve juicy segments and is rather tart. The plant reproduces well from polyembryonic seeds, each seed carrying two or more incipient plants.

The key lime originated in Southeast Asia. Arab merchants brought it to al-Andalus, where it was cultivated. Spaniards and Portuguese brought it to the New World in the early part of the sixteenth century. It then escaped back into the wild, and became naturalized in parts of the West Indies—hence the name "key lime" from the geographic keys (islands) off the coast of Florida.

Nowadays, key limes are grown in India, Mexico, Egypt, and the Caribbean, and commercialized via the juice, limeade.

Valencia oranges are one of the most widespread orange varieties, cultivated in both California and Florida in vast amounts. The fruit is sweet and very juicy. It is easy to cut and seedless.

The name is misleading, however. These oranges did not originate in Valencia, on the Mediterranean coast of Spain. The Valencia sweet orange is a native of China. Spanish or Portuguese voyagers brought it back to Europe. The Valencia cultivar that we know today is much more Portuguese than it is Spanish. In the early 1860s, the English nurseryman Thomas Rivers (1798–1877) of Sawbridgeworth imported it from the Azores and catalogued it under the name Excelsior.

Rivers and his orange crossed the Atlantic Ocean in 1870, when he provided trees to both S. B. Parsons, a Long Island nurseryman, and A. B. Chapman, of San Gabriel, California. The Florida connection came about when Parsons sold and shipped trees to E. H. Hart in Federal Point. Hart renamed the variety and commercialized it under his own name, as Hart's Tardiff, in 1877. By that time, the competition between California and Florida for control of the northeastern U.S. citrus market was already fierce. The first carload of Valencia oranges shipped to eastern markets (in the same year, 1877) came from J. R. Dobbins of San Gabriel.

In the meantime, Chapman had also renamed the fruit, but not after himself. A Spanish visitor had pronounced it similar to another late-maturing variety, from the Valencia area. And so the Valencia orange received its name at the foot of the San Gabriel Mountains.

I recommend a somewhat unusual way of enjoying Valencia oranges (which are available from February to June, if you buy from a Florida supplier):

FRIED VALENCIA ORANGES

4 large seedless oranges
4 tablespoons brown sugar
1/8 teaspoon nutmeg
1/8 teaspoon mace
1/4 teaspoon cinnamon
1 cup flour
1 1/2 teaspoon baking powder
1/4 teaspoon salt
3 tablespoons brown sugar
1 raw egg
1/3 cup milk
2 tablespoons olive oil

Garnish:
4 tablespoons mustard
4 tablespoons brown sugar

- Carefully peel the oranges and separate the segments.
- Add the sugar, nutmeg, mace, and cinnamon.
- Combine the flour, baking powder, salt, and brown sugar.
- Blend two tablespoons of oil, the egg, well beaten, and the milk.
- Thoroughly stir this liquid into the dry mixture to form a thick batter. If the batter is thin, add a little more flour; if it is too thick to evenly coat the orange segments, then dilute with more milk.
- Chill batter for 1 1/2 hours.
- Heat the remaining oil in a heavy skillet until hot, but not smoking.
- Dip orange segments in batter to coat thoroughly. Drop into hot oil and fry until nicely browned.
- Serve warm with the mustard and brown sugar in separate spice dishes.

The Jaffa sweet orange also answers to the name Shamouti. It originated as a limb sport on a tree near Jaffa, Palestine (now Israel), in 1844. The peel, thick and leathery, comes off easily. The fruit itself is very juicy and sweet. This orange quickly dominated many plantations in Lebanon and Israel, where it constitutes three-quarters of the citrus crop.

Introduced into Florida around 1883, the Jaffa made large inroads there, as well as elsewhere in the United States. But it had liabilities as well as assets: the tree tends to bear fruit only in alternate years, does not hold onto its fruit very well, and is often the victim of a fungus. Hence, Jaffas have now largely disappeared from the United States.

––––––––

The navel variety of orange is reminiscent of castrato singers—outstanding, but without progeny. Originally, the navel was a sweet orange in Portugal by the name of *Seleta*, or *Selecta*. It was brought to Brazil, where, in the state of Bahia, a chance hybridization produced a limb sport. A structure at one end of the fruit, similar in appearance to a navel, led to this novel variant being named *umbigo*, Portuguese for "navel." Outside of Brazil, it came to be known at first as *Bahia*.

From the beginning of the nineteenth century, travelers to Brazil commented on this unusual fruit: it was easy to peel, its segments separated readily, and it was juicy and sweet. Two Germans, Karl Friedrich Philipp von Martius, a physician and botanist, and his zoologist colleague, Johann Baptist von Spix, noted in 1817 the excellence of both *Seleta* and *umbigo*. A book by Domingos Rebelo in 1829 mentioned *umbigo* among the existing orange varieties. The French painter Jean Baptiste Debret, in his *Voyage pittoresque et historique au Brésil* (1839), drawing on a fifteen-year stay in Brazil, also noted the "delicate flavor and the infinitely sweet taste" of the *umbigo*.

The *umbigo* is seedless because the tree produces no pollen and few ovules. Accordingly, the tree has to be grafted onto other varieties. Its seedlessness is not the fruit's only anomaly; the so-called navel is actually a rudimentary secondary fruit nesting in the top of the primary fruit.

In the aftermath of the Gold Rush of 1849 and the Civil War, numerous Easterners settled in California. Often, they came as groups of families that together founded a colony. As a rule, such groups were united by a religious creed or some other shared ideal. One of these groups was led by John North, a militant abolitionist from upstate New York who came to California in 1870 and founded a utopian colony in the Riverside Valley.

Fellow colonists Luther and Eliza Tibbets had grown tired of the cold, rough winters in the Northeast. In 1873, Eliza wrote a letter to the Department of Agriculture in Washington, DC. She asked for advice on trees to plant in her front yard that might thrive in the California climate. She was sent three seedlings of the *umbigo* Bahia orange from Brazil. Eliza planted the trees. One was trampled by a cow, but the other two prospered. Legend has it that she used her dishwater to water them: Luther

Tibbets was too lazy or too cheap to install irrigation. One of these two trees still survives in downtown Riverside at the intersection of Magnolia and Arlington avenues.

As the story goes, Eliza served her oranges at a housewarming party, and they were an instant sensation. In any case, she started a mail-order business, selling budwood at five cents a bud, according to some sources, or up to five dollars each, according to others. Eliza Tibbets, a Queen Victoria look-alike, made money and became very influential. Her three orange trees were the foundation of citriculture in California, first in Riverside and later in the whole citrus belt extending from Pasadena to Riverside.

Eliza Tibbets shares the credit for starting navel orange cultivation in California with the man who sent her those three trees. William Saunders (1822–1900), a Scotsman and a botanist who trained at Kew Gardens, came to the United States in his mid-twenties. He owned nurseries and was a landscape architect in the Baltimore-Washington area. He designed many parks in the eastern United States, including the military cemetery at Gettysburg.

In 1862, Saunders was appointed botanist and superintendent of horticulture at the newly created U.S. Department of Agriculture. His long tenure at USDA was very active and creative. He made the nation's flora his preserve. During his lifetime, his library of books and pamphlets on agriculture and horticulture was the largest in the United States.

This immigrant saw his new country's land as a barren field to enrich with horticultural parks and plants of foreign origin. In particular, Saunders introduced the cultivation of a number of fruits from abroad. He brought in the Japanese persimmon, and he planted hardy Russian apples in the extreme north of the country.

Saunders had a seminal role in citrus cultivation. In 1869, he introduced *Poncirus trifoliata*, the Japanese hardy bitter orange, and of course, he sent those *umbigo* Bahia oranges to Eliza Tibbets.

———

Frank N. Meyer (1875–1918), whose itinerant life belongs in a movie, was born Frans Nicholas Meijer in Amsterdam, in 1875. The Amsterdam Botanical Gardens fascinated him as a boy, and he spent all his free time there. He came to the attention of the director, Hugo Marie de Vries (1848–1935), who gave the boy a job as a gardener's assistant at the age of fourteen.

Hugo de Vries was a scientist and an educator, a professor of botany at the University of Amsterdam. He is best remembered for his rediscovery, with others, of Mendelian genetics. He took an interest in Meijer, helping him learn foreign languages (English and French) and looking after his education in the sciences. Vries's clear intent was that Meijer become a scientist. All told, Meijer spent eight formative years at the Amsterdam Botanical Gardens, interrupted only by military service. He rose to become head gardener in charge of the experimental garden.

The young man had a solitary temperament, happiness seeming out of his grasp: "I am pessimistic by nature, and have not found a road which leads to relaxation. I withdraw from humanity and try to find relaxation with plants," he wrote to a friend on October 11, 1901. In search of personal equilibrium, he would travel on foot to discover foreign scenery and plants. He went through nearby Belgium into Germany, France, Switzerland, and Italy. He made a trip to Spain, where the orange groves were his main interest.

These hiking trips started a pattern, in which Frans Meijer would work in a nursery, earn some money, and then take to the road again—in search of new plants, new sights, and the satisfaction he derived from his migrant life. Thus he left for England, spent a short while in a commercial nursery in London, and resumed his travels when he sailed for New York on S.S. *Philadelphia* in October 1901. Upon his arrival, the now renamed Frank N. Meyer quickly found employment in Washington, DC, in the greenhouses of the U.S. Department of Agriculture. Of course, he did not stay put. During the next four years, he made his way through California and Mexico, also traveling to Cuba.

At the USDA, he became friends with David Fairchild (1876–1954), who, though only twenty-two, had created a department of foreign seed and plant introduction. Fairchild was also a world traveler: he would roam the planet for thirty-seven years searching for useful plants to acclimatize to the United States. Another, older botanist also became influential to Meyer: Charles Sprague Sargent (1841–1927), the son of a wealthy Boston merchant, banker, and railroad financier, looked after a private arboretum on his family estate of Holm Lea before he was appointed director of the Botanic Garden at Harvard in 1872. Fairchild and Sargent, responding to Meyer's all-too-evident wanderlust, easily convinced him to accompany them to Asia on a mission for the USDA to bring back to the United States whatever specimens they would deem of interest.

Frank Meyer started on his first trip to China by himself (ironically, Fairchild and Sargent were unable to accompany him). This was at a time when China, in the shadow of the Boxer Rebellion, fascinated Westerners. On his first expedition (1905–1908), he went from Shanghai to Hupeh, Manchuria, and back. He shipped back quite a few plants, such as *Ginkgo biloba*, the Chinese horse chestnut (*Aesculus chinensis*), the Chinese juniper tree, and two species of Chinese persimmon. A picture of Meyer at about that time shows a dapper, mustachioed man with a prematurely receding hairline. He was awed by the cornucopia of natural species: "Our short life will never be long enough to find out all about this mighty land. When I think about all these unexplored areas, I get fairly dazzled; one will never be able to cover them all. I will have to roam about in my next life," he wrote in a May 1907 letter to David Fairchild.

This letter was perceptive: Frank Meyer's life would be short indeed. He did return to China for another three plant-collecting expeditions, in 1909–1911, 1912–1915, and 1916–1918. He roamed all over that vast country, collecting tens of thousands of specimens. The number of attendant introductions of useful plants in the United States was approximately 2,500—a stupendous figure!

By 1918, however, traveling in China had become highly dangerous for any foreigner. The political situation was in turmoil. Frank Meyer wisely opted to return to the United States. He sailed down the Yangtze River to Shanghai on May 28, 1918. He was probably planning to leave Shanghai for the United States, but he never made it. He died under mysterious, suspicious circumstances during this last leg of his trip, at the age of only forty-three.

As a part of his legacy, Meyer happened to notice, in an ornamental pot near Beijing an attractive small tree bearing sweet-tasting fruit, eerily similar in outer aspect to lemons. Meyer introduced this plant into the United States as S.P.I. (USDA Shipping Point Inspection) no. 23028. And Meyer lemons, as they became known, started colonizing the backyards of many homes in California, where they found a most congenial climate.

―――――

Walter Tennyson Swingle was born on a farm in Pennsylvania on January 8, 1871. After his family resettled in Kansas, the young boy became fascinated by the local flora. From Kansas State Agricultural College he borrowed a copy of Gray's *Manual of Botany*. This led him to enroll in the college at the age of fifteen. Supervised by Professor William A. Kellerman, he trained as a mycologist. By 1891, when he joined the USDA, Swingle had authored or coauthored no fewer than twenty-seven scientific papers.

Following his arrival in Washington, DC, Swingle was sent to survey the citrus-growing areas of Florida, and his interest in citrus was thus sparked. In 1894, a freeze destroyed the citrus crop in Florida and put a brutal stop to his work there.

Swingle decided to take advantage of this setback to learn more biology. He went to one of the best laboratories in the world, that of Eduard Adolf Strasburger at the University of Bonn. There, in 1895–1896, he carried out studies on cellular structure in plants, proving the existence of the centrosome. This work earned him a master of science degree from Kansas State Agricultural College. He took another leave of absence from the USDA in 1898 and traveled to Leipzig for further study under Strasburger, who had moved there. In Europe, he fell in love with Lucie Romstaedt, his tutor in French; they were married in 1901.

During the first decade of the twentieth century, Swingle's assignment was to develop date culture in the United States. He introduced the fig wasp into the United States for pollination. By that time, his interest in mycology had waned, and his main interest had become citrus. Hired initially by the USDA to work on a citrus canker as a mycologist, he had become more interested in the trees harboring the disease than in the fungus itself.

He taught himself citrus genetics and the skills of tree husbandry— particularly the production of hybrids sporting strong points of the varieties being combined. Led by scientific curiosity, he dared to dive into Mendelian genetics as soon as Mendel's laws were rediscovered by Hugo de Vries and others at the turn of the century. He trained himself as a plant geneticist.

In 1910, Lucie Swingle died of typhoid fever. The following year, Swingle met a fellow botanist visiting Washington. She was Maude Kellerman, the daughter of his scientific mentor, William A. Kellerman. He and Maude were eventually married in 1915. From this time on, Swingle worked predominantly on dates and citrus. Numerous new varieties and cultivars were his doing, including the minneola and the tangelo.

Walter Swingle was an inveterate collector. From exotic citrus species, he moved on to exotic books. He had a major role in building the collection of Orientalia in the Library of Congress. As a scout, he found over 100,000 Chinese books on botany, which he helped the Library of Congress to acquire. He collaborated with Michael J. Hagerty, a Chinese translator for the USDA, in translating Chinese botanical papers and treatises.

———

Herbert John Webber (1865–1946) had a life not atypical for an American scientist during the first half of the twentieth century. Raised on a farm in central Iowa, he entered the nearby University of Nebraska in 1883, where, under the influence of Charles E. Bessey, he became a botanist. He went on to study (at Woods Hole, Massachusetts, and Washington University, St. Louis) cell modifications during fertilization and division.

The year 1892 was a watershed year for Webber. He was hired by the USDA and sent to Florida as an assistant to Walter T. Swingle. Webber, who had trained in pure science, was confronted there with citrus canker and other diseases. He espoused applied research on the spot and started working with citrus growers. Upon his return to the USDA laboratories in Washington, DC, in 1897, Webber switched fields. He specialized in citrus reproduction for his Ph.D. work (1900, Washington University). But the USDA shifted him to plant breeding.

With the rediscovery of Mendel's laws, Webber saw plant genetics as the common foundation of both fields in which he had been involved. This intuition put him at the leading edge of botanical research. Indeed, in 1907, Cornell University, in Ithaca, New York, chose him as professor of botany and as head of its brand-new Department of Experimental Plant Breeding.

A few years later, in 1912, Professor Webber received another call, this time from the University of California. He was appointed director of the Citrus Experiment Station, or CES, at Riverside (and dean of the graduate school of tropical agriculture there). Since 1907, the CES had occupied 23 acres on the eastern slope of Mt. Rubidoux in Riverside. It was founded at the instigation of the citrus growers, who in 1905 had asked the state legislature and the University of California for help. The CES worked on all kinds of problems: fertilizers, rootstocks, cankers and nematode pests, heating of orchards for frost protection, and new, improved varieties of citrus.

In 1913, it was decided to increase funding for the station significantly. Most of the people involved (the University of California faculty and regents, as well as legislators) pushed for relocation to the San Fernando Valley, closer to the influential, expanding city of Los Angeles. But Dr. Webber was adamantly opposed to such a move. He wanted the CES to stay in Riverside, and he dug in his heels. However, in 1917, he willingly accepted a move to new facilities, only four miles to the east, on the lower slopes of Box Springs Mountain. Webber would work there with his staff until his retirement in 1929.

Webber's strong intuition and his pragmatic approach are exemplified by an episode many years after he retired. In 1942, the Ugli fruit was sent from Jamaica to New York to be sold there for the first time. Kendal Morton bought four of the fruits from a New York vendor and sent them to

Dr. Webber in Riverside. Dr. Webber planted the seeds and obtained thir-teen seedlings. He then smelled their leaves: six had a strong mandarin odor, three had a weak mandarin scent, and four had a perfume that evoked grapefruit or sweet orange. He was not given the time to further analyze the Ugli as he had planned, but his nose had told him the correct ancestry for this hybrid!

––––––

These biographies have several aspects in common. I wish to single out one of them: the key role of government in getting large-scale economic projects started.

I want to call attention to the role of Abraham Lincoln's adminis-tration and of Congress in the year 1862. The Homestead Act, passed that year, was accompanied by the Morrill Act, which created land grant colleges in every state, devoted to the betterment of agriculture and engineering. What was to become the United States Department of Agri-culture was set up in the same year.

Only a generation later, the results of such far-sighted vision began to be realized. Thanks to Saunders's initiative, California citriculture was booming. Florida was not far behind, the work of Webber, Swingle, and others having introduced citrus varieties that thrived in that climate. Such episodes ought to be remembered. State intervention can be a good thing. Sometimes it is indispensable, as with the dams that provide irri-gation to the citrus belt in the Central Valley of California.

––––––

Here is one last example of improvement of citriculture through the ad-mittedly indirect influence of the United States federal government:

Agribusiness can make good use of science. Its practitioners can afford to hire the top scientific guns. A case in point is the Rio Red grapefruit from Texas. The fruit owes its ruby color to an abundance of lycopene, a molecule of the carotenoid family, in both the peel and the flesh of the fruit. The Rio Red variety not only is man-made—all cultivated plants are, and can be legitimately termed "genetically modified organisms"—but is also a relatively recent addition to the several thousand known varieties of citrus. Its appearance may also qualify as a Texas tall story—for one thing, there is talk in the Lone Star State of "ten-gallon grapefruit."

Texas weather was the driving force for modifying citrus trees to make them hardier. Freezes in 1949, 1951, 1962, 1983, and 1989 wiped out millions of trees. While they were at it, Texan agronomists also aimed for a redder variety of grapefruit.

Richard A. Hensz, a Texas horticulturist, visited the Brookhaven National Laboratory, in Upton, New York, on Long Island. The BNL is part of the U.S. Department of Energy, but it is a facility run by a consortium of universities in academic style. It prides itself on a number of Nobel Prize winners, predominantly in high-energy physics. The BNL has a number of accelerators producing high-energy beams of particles and radiation.

At the BNL, Hensz began to irradiate grapefruit seeds with thermal neutrons, or with X-rays, in the hope of inducing further mutations. In 1965, he produced the Star Ruby variety, released to growers in 1970. But you can't win 'em all: the Star Ruby tree, resistant to damage by frost, proved to be unusually sensitive to damage from herbicides, and it bore fruit with unpredictable regularity.

Back to the drawing board, and to the BNL, went Dr. Hensz. In 1976, he came up with yet another new variety, named Rio Red. It is "the paragon among red grapefruits" it is advertised to be, at least so far. Made available to growers in 1984, it is now grown worldwide.

In the Rio Grande Valley, groves are laid out on alluvial soils several miles north of the river, an area of highly alkaline, very sandy loam. The soil resembles that found in grapefruit-growing areas of Florida.

A Texas grower recently spoke about the prosperity of the grapefruit crops in that region. "Our trees flower in February and March," explained Paul Heller, vice president for field operations at Rio Queen Citrus, a grower in Mission, Texas. "By April the fruit is pea-sized, by May it's the size of a golf ball, and in September or October we start harvesting, and that lasts until April or May." The trees are spectacularly productive, and a step is removed from the growing process, as storage of the crops is not an issue. According to Heller, "Sometimes you have two crops on the same tree at once. You take off only a little bit at a time; the tree is by far the best place to store the fruit." The Rio Red grapefruit, developed only recently, owes its existence to the perseverance of scientists, with a little help from the government along the way.

FOUR

Nurturing Citriculture

Brought up in the fear and nurture of the Lord
BOOK OF COMMON PRAYER

The cold wave originated in the Midwest. It moved to the south so swiftly that the air in it could not warm up significantly. It hit Florida on the night of December 29, 1894. The temperature dipped below freezing, even as far south as West Palm Beach. Elsewhere (Titusville, Jacksonville, Tampa), thermometers dipped into the teens. Whenever air temperature remains below 26°F for at least three hours, the juice inside citrus fruits freezes. This meteorological fluke destroyed the entire Florida citrus crop.

Weather is capricious, and, alas, its excesses can occur in quick succession. Such a double whammy was inflicted on Florida when a still worse freeze hit the state the night of February 7, 1895. The next day, few parts of the state enjoyed above-freezing temperatures. Whereas only crops had been lost the previous December, the citrus trees themselves were killed in the second frost. The entire livelihoods of growers were likewise devastated. Following the 1894–1895 freeze, annual rail shipments of Florida oranges plunged to 150,000 crates from a high of 5.5 million crates in earlier years.

Acts of God? Only in the legalese of insurance companies. Humankind is used to meteorological excesses and the ensuing devastation. As a species, we have learned to be resilient and inventive in the face of catastrophes. Humans are extremely resourceful when it comes to crop protection, the topic of this chapter.

———

Three examples will illuminate some of our key strategies: the devising of orangeries as shelters for citrus trees in the face of an impending freeze; the use by citrus growers of so-called smudge pots for the same purpose; and the war waged against pests, inseparable from present-day agricultural practices, in the state of Florida.

Let us first take up the story in Renaissance Italy, which had seen cultivation of citrus fruits—oranges and lemons mainly—since Roman times.

In the tale I am about to tell, cold was the first aggressor. French armies were the second. French kings returned from Italy to Paris, intent upon using citrus trees brought back from Italy as emblems of their power and as status symbols for the prestige of the French court. My narrative then leaps to the present, with tourists thronging to Versailles and the Paris museums.

The age-old domestication of citrus trees by humankind has repeatedly come under attack from insects, other pests, and the climate. Italian lemon and orange groves, which had more or less thrived since Roman times, saw their extension curtailed by the advent of the so-called Little Ice Age. In western Europe it lasted from 1650 to 1850. Alpine glaciers advanced, and the people suffered from a succession of poor harvests, with the attendant famines and miseries.

Since the fruit ripens in winter, citrus crops are vulnerable to frost. The cooler weather threatened to wipe out plantations in the northern part of Italy—anywhere north of Naples. One cannot but admire the Italians. With admirable spirit—they would not give in—and with respectable ingenuity, they invented the orangery. First they erected small wooden sheds around their citrus trees. Then they built more permanent structures. Perhaps the earliest orangery was the one added in the mid-fifteenth century to the Villa Palmieri, in Tuscany.

Orangeries are buildings providing winter shelter for citrus trees. Their survival hangs on the air temperature not dropping below freezing for several hours. The northern Italian constructions combined protection against the wind and its chill factor with a southern exposure, as well as the use of materials such as stone or brick to store and reradiate solar heat. Sometimes, auxiliary heating by wood- or charcoal-burning furnaces was used.

In Italy, these constructions ensured continued life for citrus groves during the Little Ice Age as far north as Tuscany, the Piedmont, and even parts of Ticino in southern Switzerland. Quite a few remain standing today. So-called *limonaie*—the Italian term for "orangeries"—became an integral part of many a villa built in Tuscany after the sixteenth century. Large parts of the shore on some of the lakes, such as Lago de Garda and those in the Lake District between Como and Lugano or Rapallo, still display *limonaie*.

These greenhouses and their builders turned into luxury export items. When the French armies took to invading Italy regularly from the fifteenth century on, plunder became one of the dimensions of life, both for the Italians at the receiving end and for the French, who benefited at home from the Italians' involuntary largesse.

In the late fifteenth century, the king of France, Charles VIII, brought back with him Italian artisans and gardeners. They introduced the French aristocracy to Italian Renaissance gardens. The influx of talent continued with successors to Charles VIII, such as François I. Henceforth, the Loire chateaux benefited from Italian landscape architects and gardeners. During the Renaissance, these buildings included greenhouses. The French version of *limonaie* came to be known as *orangeries*.

During the seventeenth century, the Italianate influence on garden and orangery design in France was even more pronounced. When Henri de Navarre became king of France as Henri IV, he married a Medici from Florence. Maria da Medici brought with her a retinue of Italian architects, artists, and gardeners, who set about reproducing some of the Florentine marvels, such as the gardens of Florence (Boboli, Crocetta, and so on), for the benefit of the French court. One can still admire their productions in Paris at the Luxembourg Palace (which houses the French Senate) and the adjoining Luxembourg Gardens, which belonged to the queen.

––––––––

The Sun King, Louis XIV, was the grandson of Henri IV and Marie de Médicis, her frenchified name. The design of his new palace at Versailles, especially of its gardens by Le Nôtre, merged an Italian style of garden architecture with French ideas of order and rationality, a synthesis known in French history as the Classic Age.

What do the hordes of tourists visiting Versailles seek? Many visit out of a sense of obligation. Versailles is a must on the tourist's checklist. At another level, visitors are drawn there because Louis XIV was a genius of advertising. In his time, most European princely courts followed his lead and built replicas of Versailles (Schönbrunn in Vienna, Sans Souci in Potsdam, Charlottenburg in Berlin, Hampton Court near London, and so on). Even three full centuries after his death, his reputation still draws the crowds.

At still another level, the sightseers—the targets of modern advertising, who may or may not be aware that predominantly American funding makes possible the major effort at restoring and refurnishing Versailles— are lured inside the palace. They admire the royal apartments. Little do they imagine what the place *really* looked like in the time of the Sun King—a mix between an American political convention and the San Firmin fiesta in Pamplona, with, on any day, about 30,000 courtiers milling around, eating and sleeping, fighting, gossiping, showing off, whoring, urinating, and defecating on the rugs—and out the windows—and, in general, spending most of their time just being idle—which, indeed, was exactly the King's intent.

To return to the present: tourists at Versailles flock to ticket booths, then invade the inside of the palace. In so doing, they miss the whole point: Versailles was built primarily as a garden. The originality of Versailles is in the park—to which, furthermore, admission is free. Accordingly—and paradoxically—few tourists bother taking more than a few perfunctory steps outside.

What, then, is so remarkable about these gardens at Versailles? What do they show? These gardens, the epitome of baroque, were designed as a transitional territory. They consist not only of *allées* and beds of flowers, topiaries, and shrubbery, but also, and very conspicuously, large pools. Bringing water to Versailles was quite a feat. The attempted construction of an aqueduct at Maintenon mobilized an army of workers, but to no avail. It was a failure on a grand scale. But the huge water-elevating machine at Marly, still operational today, answered the huge needs of the Sun King: the gardens at Versailles absorbed considerably more water, by one or two orders of magnitude, than the whole city of Paris.

These gardens project an image of the absolute monarch. The planners prided themselves on the domestication of trees. Many trees are severely pruned. Furthermore, trees are planted along straight lines or in other geometrical shapes.

The gardens at Versailles also pay tribute to Greek mythology, by means of sculptures that set a mini-territory around each deity, so that the stroller has the illusion of being outside of time and history, in an eternal present. Among the gods of Greco-Roman mythology, Louis XIV identified himself with Apollo, the sun god. He also felt a special kinship with the god of war, Mars. He was indeed very fond of waging war. The park at Versailles is, in fact, all about Mars. Prior to becoming the god of war, Mars had been an agricultural deity presiding over the in-between territory, the no-man's-land separating cultivated fields from the wilderness. Thus, the gardens at Versailles, which the royal hunt would gallivant through (and the King loved hunting), represented this buffer zone between civilization at its most refined and the wilderness of nature. During the seventeenth century, the French mind conceived nature as a dark, primeval forest, the location of nightmares and the realm of childhood fears, which fairy tales (such as Perrault's) express so vividly. The gardens kept this wild and dangerous nature in check. Nature contains no evil, and yet the seventeenth-century mind saw nature as evil: the better to dominate, colonize, and exploit it in distant foreign territories. The hedges at Versailles, meant to contain the wilderness beyond, portend the later Maginot Line, designed to stem the tide of Germanic barbarians beyond.

———

The centerpiece, the masterpiece of Versailles, is the Orangerie. The word "orangerie," in its Versailles context, has two distinct meanings, one for the building, the other for the gardens. The building is a magnificent piece of functionalist architecture. A long nave built in 1685 by Jules Hardouin-Mansart, with a semicircular vault, it is well lit and kept warm by a succession of large double-paned windows. The garden is an Asian-style oasis. A large circular pool of water is surrounded by a host of non-native trees, palm trees and orange trees in particular, grown in large wooden boxes. During summer, from May to October, the trees are outside. During winter, from October to May, they are brought inside the building, where the constant temperature (41°F–43°F, achieved without any artificial heat) protects them from frost.

Like other trees in the park, the decorative trees in the Orangerie are set in rows on two sides, the outside alignment and the inside alignment. In both, the trees stand like soldiers on parade, clad in their uniform of green, bearing their fruits as so many medals pinned on a chest, standing at attention while awaiting inspection by the dignitaries. The domesticated trees, especially in the Orangerie, include species quite foreign to France and its climate. The Sun King displayed them as testimony to his godlike dominion over nature.

Among them, the orange trees, which give their name to the part of the garden where they stand, owe their special status—their primacy—to their solar nature. Louis loved orange trees. He had ransacked the several hundred that Nicolas Fouquet, the superintendent of finances whom he had had jailed for graft (and for life) when he assumed real power at the age of twenty-two, had proudly displayed in his Vaux-le-Vicomte castle. He had them placed in silver tubs inside the palace, in the Galerie des Glaces.

In sum, the Orangerie at Versailles was a central piece in the representation of royal power—to itself, to courtiers, to foreign diplomats and distinguished visitors, and to the world at large.

_____

And how did it fare after the passing of the Sun King and after that of the French monarchy? Marc Bloch contrasted *seigneurie française* and *manoir anglais* in one of his justly well-known books. Indeed, the orangeries survived the revolutions of the eighteenth century better in English manors than in French castles. The proletarian French Revolution brought about the demise of quite a few of the princely orangeries, often turned into barns or stables by their new owners.

A century later, around the turn of the twentieth century, the problem throughout Europe was what to do with such lavish architectural spaces. Displaying orange trees as part of the holdings of a wealthy aristocratic collector was no longer in style—nor was it sound thinking, taxwise. No one solution has prevailed. These large structures have been put to a variety of uses. For instance, the orangery at Kew Gardens, near London by the Thames River, was turned into a restaurant. The one at Schönbrunn Palace, in Vienna, now serves as a locale for concerts. Indeed, quite a few of these constructions now shelter the arts rather than plants. Perhaps the best known, the Orangerie on the edge of the Tuileries Gardens by the Place de la Concorde in Paris, facing the twin Jeu de Paume building, has displayed to good effect, for more than half a century, Claude Monet's *Nymphéas*.

_____

Let me throw in a strange idea, both irrational and unprovable. The proposition is that nature carries culture with it. The seeds of a plant somehow ferry, in their genes, as it were, aspects of a civilization that deemed that particular plant important. I shall assert more specifically—although I won't prove it, since such a proposition is devoid of any factual support—that orange trees connect the French Sun King, Louis XIV, to the emperor of China across the centuries.

Obviously, there has been no such direct connection. If there had been, it would have followed the long chain of East-West trade along the Spice Route, earlier evoked as the path of the gradual horticultural acclimatizations in the various intervening countries. However improbable, this remote influence may have some rhetorical value nevertheless. Could Marco Polo have told a wealthy Italian patron in confidence that the emperor of China safeguarded his orange trees in a special building during winter? Could a copyist of an Arabic manuscript in a medieval abbey, such as the one at Monte Cassino, have shared the idea with fellow monks, one of whom was trusted confessor and confidant to one of the Italian *condotierre*? Was it a discarded scrap from Thomas More's *Utopia* that somehow made the rounds of gardeners? Many such scenarios can be devised for the transmission of the idea of royal magnificence, as embodied in rows of orange trees, from capital cities in Asia to western Europe during the late Middle Ages. To give body to this hypothesis, the reader might see it as the seed of the plot for a novel, in the tradition of *The Name of the Rose* or *The Da Vinci Code*.

————

It does not follow from the destruction of buildings such as orangeries that they had no lasting cultural imprint. In the collective psyche, citrus fruit remained durably associated with the lifestyle of the aristocracy. One of those enduring traces is the devising by a Parisian chef, officiating at La Tour d'Argent, of *le canard à l'orange*. This dish achieved mythical status, not only because of the cost of dinner in this restaurant, but also because of an advertising ploy of genius: the assignment of a number to each Challans duck served. If you refer to the Web site for La Tour d'Argent, you will see that this dish has indeed achieved the highest gastronomical status by having been served to royalty. Edward VII enjoyed no. 328 in 1890, Alfonso XIII of Spain ate no. 40,312 in 1914, and Emperor Showa, otherwise known as Hirohito, was served no. 53,211 in 1921.

————

Were egalitarian Americans immune to the aristocratic associations of citrus trees and fruit? Not completely. This link came from France, together with chefs and their gourmet preparations. Julia Child was responsible, almost single-handedly, for a revolution in taste when she brought a French influence to American kitchens with the publication of her book, *Mastering the Art of French Cooking,* at the beginning of the 1960s. An influx of French chefs followed. For instance, Jean-Pierre Goyenvalle would reign supreme over the Washington, DC restaurant world with his Le Lion d'Or from the mid-1970s until his retirement at the end of the twentieth century. He was acclaimed in particular for his orange soufflé, a dessert that Julia Child had him prepare on her PBS series.

———

But orange soufflés were a long way off when citriculture began in the United States. Its first major transformation, from cottage industry to major economic player in both California and Florida, came during the last decades of the nineteenth century. Citrus growers, otherwise powerless in the face of the railroad companies and the rates they were charging to transport crates of citrus fruit to the large urban centers in the Northeast, organized themselves into cooperatives, which gave them bargaining power.

The cooperatives gave the growers more control over the price of citrus fruit, which also allowed them to advertise. Their advertising sold not only citrus fruit, but also California as an Edenic state of permanent sunshine and leisurely living. Similar trends affected Florida at about the same time. In both states, citriculture started to coexist with the influxes of people it had indirectly contributed to attracting: tourists and retirees started to flock to both sunshine states. As is the rule with many social fashions, the wealthy initiated the moves, and the middle class followed, in much greater numbers. This trend in turn led to the mass consumption and the mass production of citrus fruit in both California and Florida.

———

Let us now reflect on the vulnerability of citrus crops. Citrus trees have become acclimatized in both California and Florida. First brought there by the Spanish, they were planted at great distances from their geographic area of origin—China or India, depending on the variety. Like Italian or Spanish plantations, American citriculture became hostage to the vagaries of climate as well as diseases—all the more a threat once citrus became a monoculture in Florida.

Consider the effects of catastrophic drops in temperature. A "minor freeze" occurs when temperatures fall below 28°F for at least four hours. A minor freeze affects only the fruit on the tree. Florida has suffered many minor freezes since the nineteenth century. An "impact freeze" occurs when temperatures drop well below 28°F for long periods, which kills trees. The state of Florida has experienced five impact freezes. Each time, citrus growers have responded by moving farther south. The first two impact freezes occurred in 1894 and 1895, while the third—on December 12 and 13, 1962—owed its severity to the mildness of the preceding weather. Citrus trees react to cold weather by entering a semidormant state, which makes them tolerant of an occasional freeze. Temperatures as low as 15°F killed more than 1.6 million trees in the northern part of the Florida citrus belt—Marion, Lake, and Volusia counties. Other parts of Florida lost more than 500,000 trees.

How were citrus fruit and trees protected from an impending freeze in twentieth-century America? In that democratic society, in which citrus groves were a livelihood rather than a luxury, the solutions were smart, quick, low-cost and low-technology fixes. Basically, two kinds of approaches were used: heating the air or shielding the plants from the icy atmosphere by spraying them with water.

My wife was born in Los Angeles. On weekends, her grandparents would drive to the mountains, where they owned a cabin, and would sometimes bring her along. The road passed by citrus groves. She has a vivid memory, although she was a young child at the time, of smudge pots. These were large canisters a couple of feet tall, filled with oil and other flammables, such as pieces of old tires. Narrow holes or slits let in only a little air. They burned during cold nights, emitting a thick, black smoke. Witnesses remember how, at dawn, they would totally obscure the clear sky. That was the intent. When frost threatened the orchards on a clear night, the blanket of smoke from the aptly named smudge pots insulated the atmosphere above and around the trees from ominous overlying, colder layers.

A side effect of the smoke from smudge pots was air pollution, which affected the whole Los Angeles metropolitan area—a dress rehearsal for the air pollution and smog from automobile exhausts that plagues the area today. There were protests, and city authorities banned the use of smudge pots around 1950—approximately the time of "the shift," when most citrus growers left the Pasadena-Riverside citrus belt for a new one in the Central Valley of California.

The other approach to protecting citrus trees threatened by a freeze—not much of an option in the desert environment of Southern California—is spraying them with water. This technique takes advantage of two scientific facts. The first is a principle taught to schoolchildren: the temperature for any change of state is a constant, and it remains so for the duration of the change. The Celsius scale is anchored to this principle: zero is set as the temperature for the melting of ice; 100°C is the temperature at which water vaporizes. As water undergoes the transition from its liquid to its solid state, its temperature will remain 0°C as long as there remains even a single droplet of water in coexistence with the ice. Its temperature will dip below freezing only if liquid water is no longer present. Accordingly, preventing a freeze could not be simpler: keep dousing or spraying the trees with liquid water as long as low temperatures threaten.

Farmers in areas throughout the world where frost can destroy their crops know this watering ploy. It is resorted to, for example, by wine growers in the Valais canton of Switzerland, which produces Dole and Fendant wines. And it is routinely used by citrus growers in the Central Valley of California and in Florida whenever a frost has been forecast.

Water is a most unusual substance; in particular, it shows a peerless ability to ferry calories. Among all chemicals, water stands out as an efficient heat-carrying fluid. For evidence of this second scientific fact, one need only think of central heating by water-heated radiators or the cooling of nuclear reactors in power plants with water from a nearby river. In the same way that liquid water transports heat from a hot source (such as a heater) to a cooler place (such as a bathtub), it is able to keep a citrus tree warm enough so that it won't freeze. It is able to evacuate the cold from the tree, as it were. It is further able, if worse comes to worst, to solidify into ice.

The massive spraying of an orange grove in the frigid night air is quite a spectacle. Sometimes the temperature drops so low that the high-pressure water jets coming out of a nozzle will freeze almost instantly in the air. The machinery involved looks like a snowmaker, which ski resorts use to make artificial snow. Indeed, there is a story of the first snow-making machines being set up at Bromley, Vermont, after the owner and developer of this ski resort, Fred Pabst (of the Milwaukee beer-brewing family), saw the spraying of citrus orchards during wintry weather on a trip to Florida. The story is probably apocryphal. True, Pabst installed his first rope tow at Bromley in 1939. I don't know when the first snow-making machine was installed on the slopes at Bromley, but I suspect it was after Wayne Pierce had devised them in 1950.

─────

Drought is another menace to citrus trees. It is very much a fact of life that citrus trees convert water into gold, but they cannot thrive in the absence of water. The Arabs realized this during medieval times. Showing prodigious technological prowess, they installed the sophisticated irrigation networks in al-Andalus that we have already described. Similarly, Southern California was able to turn citriculture from a profitable venture into the agribusiness it became at the end of the nineteenth century only through irrigation. Engineers such as George Chaffey introduced irrigation systems on a large scale, which allowed communities such as Etiwanda, Ontario, and Pasadena to develop. When citrus growers in Southern California were forced to relocate to the Central Valley, irrigation alone made it possible. Their good fortune was a consequence of the huge public works projects of Franklin D. Roosevelt's New Deal. California citrus growers were direct beneficiaries of a Keynesian economic program. Citrus groves converted federal government capital investments of an unprecedented magnitude into income for the individual growers.

———

Citrus trees are liable to suffer from epidemics, another great vulnerability, and nowhere as much as in monoculture systems, such as those found on large tracts of land in the state of São Paulo, Brazil, and in Florida.

Growers obviously want to protect their citrus plants because they represent income—big income. In Florida, citriculture earns $9 billion yearly. Ninety-two million Florida citrus trees produce three-quarters of the citrus fruit consumed in the United States. Apart from big chills, pests are the other big threat. There are numerous ills afflicting citrus trees, and those I choose to present here are only illustrative.

The first has a name out of a Walt Disney cartoon: the broad-nosed weevil. Popularly known as a bug, this insect belongs to the coleopter group. It is considerably more beautiful than its ridiculous-sounding name would suggest. Known to entomologists as *Diaprepes abbreviatus*, it is in fact a most handsome insect. The adult looks like a Jaguar car, with a conical thorax, a head shaped like the re-entry cone of a rocket, and, rearward, an acorn-shaped abdomen. The gorgeous bodywork is inscribed with long dark streaks on the cuticle. And the head sports, way up front, a superb pair of antennae.

Adult weevils have a life expectancy of about four months. The female lays up to five thousand eggs in the relative security of the space in between citrus leaves, where the aggregated eggs look like the spread on a miniature sandwich. The wormlike larva drops from this aerial nest onto the ground. There it burrows and heads for the roots of the tree, wherein lies the threat. A few larvae are enough to kill a mature tree by chewing at its structural roots. In the Caribbean, this particular pest destroys on the order of $100 million worth of citrus crops every year. In victimized groves, the rate of infestation is about one new weevil per tree and per week.

All sorts of weaponry have been tried against the broad-nosed weevil. So far, none has been fully effective. One tactic is to spray threatened citrus trees with suspensions of kaolinite (China clay) in water. When the leaves become coated with such a film, the female weevil cannot lay her eggs. In addition, the film is impermeable to the aromatic molecules emanating from the plant. The weevils ordinarily pick up these chemical cues with their antennae and use them to locate the plant and latch onto it. Kaolin also reduces damage to leaves from adult weevils by about three-quarters.

Aphids stand out among other citrus-attacking insects. For instance, the brown citrus aphid, *Toxoptera citricida*, feeds on new citrus growth. This aphid sucks the sap of the tree and directly weakens the plant.

Some aphid species are even more dangerous because of the *tristeza* virus they carry. This is especially true of *Aphis gossypii*, the cotton aphid; of *Spirea*; and of the brown aphid. All citrus trees cultivated from sour orange rootstock, which includes a high percentage of all the domesticated varieties, are susceptible to this virus.

And then there is canker. The very name of this disease, with its echo of cancer, strikes fear into the hearts of citrus growers. It is a bacterial infection by *Xanthomonas axonopodis*, of which there are three strains, spread by wind-driven rains. One of those strains, the Asiatic citrus canker, has infected about a million trees in Florida.

Canker shows up as brown and yellow spots on citrus leaves and on the rind of the fruit. And that is all. Nontoxic to humans, it does not affect the taste of the fruit, nor its maturation. It is ugly, period. Only the outward appearance of the fruit is altered.

Citrus canker thus might be just an external symptom of the striving for perfection endemic to American society. The consumer is king, and he is spoiled. American corporations endeavor to provide only the best products. Destruction of canker-affected trees and fruit costs growers in Florida an estimated $350 million per year, representing about 4 percent of the yearly income from citrus.

———

There are yet other problems in citriculture. The ripe fruit is fragile. If it falls to the ground, it spoils. It has to be handpicked from the tree. The twentieth century witnessed symbolic changes in the delivery of citrus fruit to consumers. From luxury items, individually wrapped in crates, they have turned into a commodity, with orange juice transported by tankers in tens of kilotons.

Which brings up the cozy relationship between citrus and capitalism. The ills that periodically beset citriculture have enabled a few people to step in and take control. As a rule, empires are born in times of strife. This is true of economic empires too. Thus it started for José Cutrale, Jr., the Brazilian emperor of citrus, as we shall see. In a similar manner, the Florida citrus barons, people such as Charles von Maxcy (killed by a hired thug in 1966), Jack Monteith Berry, John V. d'Albora, and Ben Hill Griffin, Jr., were quick to take advantage of catastrophes, such as impact freezes, to purchase land at low cost and increase the area of their citrus groves. Griffin, for instance, owned nearly 100,000 acres and controlled about 275,000 through private holdings. Following the devastating 1962 impact freeze that wiped out much of Florida's citrus production, Jack Berry—against everyone's advice—took a chance and planted 6,000 acres near La Belle. In 1982, once again after a devastating freeze, Berry increased his citrus and land holdings by adding 16,000 acres in Sarasota County, paying $25 million cash.

By the 1980s, these citrus barons had come to dominate the citrus agribusiness. But by the turn of the twenty-first century, they had given way to multinationals such as Sucocitrico, the holding of the Cutrale family of Brazil. One of the reasons is that only multinational producers have enough negotiating power to do business with multinational distributors, such as Coca-Cola. The current tendency is for huge financial funds to take over the juicy citrus business, Louis Dreyfus being one of them.

And thus the Jeffersonian dream came to an end. Especially in the West, Thomas Jefferson had willed the building of a society of equals, prosperous from cultivation of the land, but without any possibility of a pyramid structure with only a few wealthy people at the apex. The first decades of citriculture in California embodied this dream, with settlers forming communities in the first citrus belt. When the growers moved to the Central Valley after World War II, the federal government greatly helped them by providing the necessary water. But corporations took over afterward, in both California and Florida.

———

When economic forces are involved in the effort to protect citrus crops, things get complicated. In the year 2002, facing an outbreak of citrus canker in Florida, Governor Jeb Bush made an executive decision to destroy diseased trees. Thus, 1.5 million trees in commercial groves were sacrificed. In addition—you can imagine the outcry—another 603,000 trees were destroyed in the backyards of about 250,000 homeowners.

This poses the dilemma, familiar to political theorists, between the prerogatives of the state and the freedom of the individual that is familiar from issues such as the wearing of safety belts or cigarette smoking. Yet Governor Bush's decision seems a little different. He was invading the privacy of a relatively small fraction of the Florida electorate in order to preserve the well-being of a politically better organized segment of the population: the citrus growers to whom the state of Florida owes a substantial part of its prosperity (its other main economic resource being tourism).

The American people hold the pursuit of happiness as a worthy goal. In this particular instance, Michael Oakeshott's views are applicable: in a civil society, as exemplified here by the state of Florida, the rule of law ought not to be used for the hypocritical promotion of general welfare (while actually preserving the wealth of a few politically influential individuals), nor for the protection of any other, similar abstraction. The law should not reach beyond arbitration between the interests of rival groups—in this case, the citrus growers, on one hand, and on the other hand, private citizens growing citrus trees in their backyards.

———

The domestication of nature will always remain, in the strongest sense, an adventure. Nature may at any time reclaim its dominion by means of a devastating freeze or the ravages of a bacterium, a fungus, or a parasitic herbivore. Our human conquests remain fragile.

Of course, human greed keeps them alive. The next chapter will describe a few historic and geographic examples of great wealth attained through cultivation of citrus.

PART TWO

Mining Value from Citrus

California Dreamin'

The planting of citrus groves in Southern California—the subject, for the most part, of this chapter—is a tale of American religion and Jeffersonian ideology, of technology (mainly the railroads), and of advertising, whose wizardry creates mental images that it turns into cash. The tale also features the shift of the California citrus belt from the Los Angeles area to the Central Valley. It ends with a similar story from New Zealand, a long way off.

I introduce the religious aspects of this tale with a note on Portugal: even though the climate of that country is well suited to growing citrus fruit, it did not make for cash crops, a seeming paradox.

Why go beyond self-sufficiency, the Portuguese seem to have asked themselves. Their limited production of citrus fruit gives the impression that the country deliberately refrains from international trade. The entire amount of citrus fruit produced there is consumed there. There are no exports to speak of, in marked contrast with neighboring Spain.

The diverse attitudes toward trade of the Portuguese, Spaniards, Italians, and French—all Mediterranean peoples with a Latin-based language—notwithstanding, it is noteworthy that two Portuguese products, port and citrus, provide a textbook example of the heavy influence of religion on economics.

Max Weber's influential thesis, later reiterated by R. H. Tawney and many others, held that the Reformation bred a Protestant ethic that, in contrast to the Catholic tradition, encouraged the making of personal wealth from hard work. According to this thesis, the Protestant notion of a moral life connects directly with the rise of capitalism.

Jumping now from the early modern period to the nineteenth and twentieth centuries: while the Protestant religion provided a positive incentive to Anglo-Saxon merchants, harsh memories from the Inquisition were deterrents to would-be Portuguese traders. Their ancestors had been hounded by the Inquisition at the beginning of the sixteenth century. In the Algarve, for instance—the southernmost part of Portugal—inquisitors were diligent in ferreting out Jews and in persecuting converted Jews, the so-called New Christians. Merchants were a prime target for these Catholic, anti-Semitic prosecutors because commerce had previously been the province of Jews. Thus, in the Algarve, where this persecution is amply documented, as well as in the rest of Portugal, the Portuguese abandoned the mercantile professions. The English stepped in to replace them.

This history is still apparent today, as a look at the label of a bottle of port will tell you. Some years back, while in Lisbon, I sought the advice of a friend, Professor Jorge Calado, as to which brand of port I should buy. His advice was simple: go for the British names. Indeed, about half the well-known and well-established companies in the city of Porto responsible for the trade and export of port wines—which traditionally float down the Douro River on special barges from the production areas upstream—were initially family businesses owned by British merchants. The names (readers might recognize a few) include Cockburn, Taylor's, Sandeman, Osborne, Dow's, William Pickering, Sainsbury's, Harrington's, Harvey Nichol's, Whigham's, Churchill's, Jenning's, Whitwsham's, John Grape, Croft, and House of Commons (for the exclusive use of its members).

Citrus offers an exact parallel. In the century from 1780 to 1880, the island of São Miguel in the Azores—a Portuguese territory to this day—produced lemons and oranges predominantly for the British market. In 1860, Portuguese oranges accounted for 74 percent of the total imported into London, Liverpool, Hull, and Bristol. More than half came from São Miguel. The exporters worked out of Ponta Delgada, the main city on that island. But who were they?

This particular trade started in 1751 with a Portuguese, João Simas Camelo, who sent a shipment of 3 1/3 large crates of oranges to Cork, in Ireland. But it later became almost entirely held in the hands of foreigners. By 1820, there were nine exporters operating out of Ponta Delgada. Only one of these was Portuguese, Jacinto Ignacio Rodrigues Silveira. Another was a Prussian, named Sholtze. Thomas Hickling was originally an American. All the others were British: Ivens & Burnett, Brander, Nesbitt, William Harding Read, Cockburn, and Lesbie. A dozen years later, the situation was the same: the highly lucrative trade in São Miguel oranges was almost entirely controlled by foreigners.

Thus, the fear of persecution during the Inquisition remained as a powerful undercurrent even into the nineteenth century. And at the beginning of the twenty-first century, it still endures: the millennium-old prohibition against Christians making money from banking or trade is very much alive in Portugal, a Catholic country where the Church retains a huge influence. It goes a long way toward explaining why Portugal, with a perfect climate for growing citrus fruit, exports so little of its production.

The restraint of the Portuguese is in stark contrast to the wealth that Anglo-Saxons—not only the British, but also the Americans—made from citriculture. I shall focus here on California rather than on Florida, although these two citrus-producing regions have similar histories. The California story, because of the climate in Southern California, is centered on Los Angeles: a line from the Mamas and the Papas song, "California Dreamin'," "I'd be safe and warm if I was in L.A.," speaks to the power of the image of the state as a temperate Eden.

The factors determining the citrus belt in Southern California were sun, mild temperatures, almost no frost, moist air, and a porous alluvial soil. The proximity of the orange groves to foothills and mountains— which one sees depicted on many crate labels designed between 1885 and 1945—was another favorable factor. The mountain ranges provided both a windbreak and a reservoir of water.

The expansions of Los Angeles and Miami, as tourist destinations and as desirable locations for settlers, occurred at about the same time, during the second half of the nineteenth century. Both came about because of the arrival of railroads. Both regions advertised themselves effectively using citrus fruit at a time when the fruit was symbolic of exotic lands of sun and leisure. Orange blossoms; palms; olive trees; a cornucopia of exotic species, aromatic fragrances, and luscious scents: such cues come straight out of the Bible! Would-be settlers in the northeastern United States took them as referents to a new Promised Land in the West or the South.

Citrus fruit was arguably the best possible advertising for the wonders of California and Florida. Orange trees were then romanticized, standing as an alluring feature in exotic landscapes of the tropics, as in these lines from Tennyson, written in 1847:

O Love what hours were thine and mine
In lands of palm and Southern pine;
In lands of palm, of orange-blossom,
Of olive, aloe, and maize and vine.
—*The Daisy*, 1847, i

The cultural heritage of both Los Angeles and Miami was Christian and Hispanic, especially so in California, where El Camino Real, the royal road from Mexico to San Francisco, interconnected Spanish Catholic missions set up along the coastline.

———

Los Angeles had already undergone a few metamorphoses since Spanish times. It was at first, and until the 1840s, a Hispano-Mexican colonial village of large estates (*ranchos*) largely devoted to raising cattle. In the 1850s, it became a conglomeration of vineyards and small orange orchards as the former *ranchos* were divided into ten- and twenty-acre farms. After the switch from cattle ranching, the henceforth agricultural Los Angeles County continued its slow growth until 1876, when the arrival of the Southern Pacific Railroad connected Los Angeles to the eastern United States via St. Louis.

California received an initial massive influx of immigrants after the Gold Rush of 1849. Soon after, the transcontinental railroad opened up the West to the demographic influx from the East. Just a year after its arrival in Los Angeles, in 1877, a citrus grower, Joseph Wolfskill, sent a railcar filled with navel oranges to St. Louis. They arrived in good shape, giving rise to good press for sunny Southern California, a land of bountiful agricultural crops.

A steady supply of oranges to the eastern U.S. market was quick to follow. Tourists came to visit Southern California in droves. Newcomers were attracted by the agricultural bounty advertised in the eastern United States by the arrival of carloads of California navel and Seville oranges. A panic in 1885, caused by a blight that affected the citrus plantations, led to the real estate boom of 1885–1888: owners of orange groves sold out to real estate investors under the pressure of the numerous immigrants flocking to California. Los Angeles became a budding metropolis; its population tripled in the 1880s and then doubled again in the 1890s. Orange groves ushered in the phenomenal urban sprawl of Los Angeles.

The Los Angeles story has several threads. One is the entrepreneurial audacity of the Wolfskill family. Another is the commercial rivalry between the various railroads. A third is the acclimatization to Southern California of varieties of citrus, such as the navel, that had not been grown there before and which thrived in the climate. A fourth is the series of booms and busts as one speculation followed another, at the pace of about one a decade, during the second half of the nineteenth century: the Gold Rush; cattle breeding; sheep raising; citrus production; real estate. Yet another thread is the violence to which California is prone. During those pioneering times, the ferocity of nature (earthquakes, torrential rains, droughts) was matched only by the often ferocious human behavior in outposts of the Wild West.

———

But let us return to the Wolfskills. In 1841, when California was still a Mexican possession, the cross-continental wagon train of the Workman-Rowland party arrived in the Los Angeles area from Pennsylvania. In that same year, William Wolfskill, one of its members, planted the first commercial orange grove in California, two acres in area. By 1855 or so, he had planted another two thousand trees to enlarge his orchards. By 1860, Wolfskill's orchards extended to more than a hundred acres planted with orange trees. A couple of years later, the Wolfskill property contained two thirds of the new state's citrus trees. In 1866, William died, and his son Joseph Wolfskill took charge of the family orchards.

The Santa Fe Railroad arrived in Los Angeles in 1885. Competition between it and the Southern Pacific Railroad immediately led to a fare war, and the cost of train transportation dropped significantly. In the following year, 1886, the Wolfskill orchards sent the first *train* loaded exclusively with oranges to the eastern markets. Each crate of oranges reaching people in the East was an inducement to come West. The growers came up with the slogan, "Oranges for Health, California for Wealth." By 1887, the Southern Pacific Railroad was bringing 120,000 visitors per year to Los Angeles. Furthermore, even given the stranglehold that the railroads had on the citrus producers, the cost of shipping oranges gradually decreased from three cents to one cent per ton per mile.

———

Which varieties of oranges were being grown for shipment to the eastern United States? When Eliza and Luther Tibbetts planted the first navel orange trees in Los Angeles County in 1873, their seedlings came from Brazil via Washington, DC. Navel oranges, a winter-ripening crop, thrived in California, where they found the perfect climate (they tend to grow too big and not as tasty in Brazil, where the weather is warmer and more humid). During the period 1873–1888, more than a million navel orange trees were planted in the Los Angeles area. By 1880, there were 200,000 orange trees in Los Angeles County, more than five times the 1870 count. In 1889, more than 13,000 acres of Los Angeles County were planted in citrus. Shortly after the first planting of navel orange trees, the first shipment of Valencia orange trees reached Los Angeles in 1876 (imported by A. B. Chapman from the Azores). Production of this fruit, a summer-ripening crop, thus started in the late 1870s. Henceforth, Southern California could supply oranges year-round.

––––––

This tale, as recounted so far, creates a misleading sense of progress and harmony. It would seem that a single factor, the profitability of citrus crops, made California thrive and made the city of Los Angeles grow as it started exporting oranges to the big markets in the eastern United States. But, of course, history does not behave this simply. The growth of citriculture in Southern California had disasters of its own.

Southern California enjoyed a rather dormant status during the colonial era, as a Spanish and then a Mexican outpost along El Camino Real. The first major disruption to its economy and social organization coincided with the military takeover of the region by the United States (1846–1848). It came about as a consequence of the 1849 Gold Rush. The Forty-Niners who had flocked to the San Francisco and Sacramento areas clamored to be fed. Beef from the *ranchos* in and around Los Angeles became a valuable commodity. A boom ensued, and profits increased tenfold. The good times lasted a little over ten years, just long enough for people to become dependent on this source of income as if it were God-given and permanent.

But nature unshackled itself, shrugging off the new beef-raising economy. In 1861, California received fifty inches of torrential rains, which caused massive and extensive flooding. Seemingly as an instantaneous, abrupt correction of this deluge, a drought ensued. 1862 was a dry year. So was 1863. And 1864. And 1865. After these rainless years, 80 percent of the cattle were dead: the end of an era had been reached.

Vineyards were another traditional resource whose growth was spurred by the Gold Rush. By 1870, one sixth of American wine production came from small farms carved out of the much bigger *ranchos*. Los Angeles County had more than 5 million vines in 1880. But in the early 1880s, a blight attacked the vineyards, which were already facing competition from wine growers in northern California and suffering from the stiff transportation prices exacted by the railroads.

In 1880, Los Angeles County was still heavily agricultural. Orchards and vineyards dominated the southern parts of the city between the river and San Pedro Street. With 200,000 trees in Los Angeles County, citrus orchards were everywhere.

Orange groves provided a handsome income. On average, each acre in the Wolfskill orchards brought in $1,000 per year. Such bounty led to a boom in citrus tree plantations. For instance, 1,300 acres were planted in the San Gabriel Valley in 1877, increasing to 2,200 acres in 1879.

Three factors came to check the prosperity of the orange groves. The first was the avidity of Easterners for land in Los Angeles, which led to a real estate boom in 1885–1888; the price of land increased fivefold during a single year, 1887. Second, the producers—victims of their own greed and of the resulting oversupply—lost control to middlemen and to the railroads. The third factor was a pest: an insect popularly known as white scale (*Icerya purchasi*), which had been inadvertently introduced in the 1860s with nursery stock imported from Australia. "[Trees are] literally white with the voracious and virile insects in all stages of development, every leaf, limb and twig being coated completely," wrote A. F. Kercheval to the *Los Angeles Times* (October 4 1888, p. 6). The white scale was not the first blight to affect orange trees, which had also suffered from red scale and black scale. The white scale was checked only at the end of the 1880s, when D. W. Coquillett imported a natural predator from Australia, the ladybug *Vedolia cardinalis*.

By the end of the 1880s, the citrus industry was in ruins. Many citrus growers became discouraged and sold out to investors in real estate. In the 1890s, the Wolfskill grove disappeared. In 1888, part of it had been sold for Southern Pacific Railroad's Arcade depot. The rest was carved into commercial and residential building lots.

————

The economic ills plaguing the citrus industry were remedied only at the turn of the twentieth century, when the producers organized themselves into cooperatives. In 1893, T. H. B. Chamblen and P. J. Dreher incorporated all the major Southern California growers into the Southern California Fruit Exchange. It was renamed the California Fruit Growers Exchange in 1905. The latter organization was responsible for introducing the Sunkist label.

———

The period 1850–1890 was one of profound social change. In 1850, California gained U.S. statehood and the City of Los Angeles was incorporated. Starting in 1851, a U.S. Land Commission gave a legal varnish to the dispossession of previous landowners with a revision of the earlier Spanish and Mexican land grants. Enterprising Yankees were elbowing out the settlers of Mexican origin.

Los Angeles turned into a rowdy city replete with rascals and ruffians, whorehouses and gambling dens. The rampant lawlessness caused Angelinos to be nicknamed "Los Diablos" (The Devils) and led to the self-appointment in 1853 of a Vigilance Committee, the Los Angeles Rangers. During the next couple of years, the Rangers summarily executed twenty-two men. In 1855, Mayor Stephen Foster temporarily resigned so that he could take part in a lynching. In 1857, James R. Barton, then the Los Angeles sheriff, was killed, and vigilantes went on a punitive rampage. The last execution by the Los Angeles Rangers was recorded in 1870. In October 1871, an anti-Chinese pogrom occurred.

Meanwhile, the prosperity accruing from the orange groves, the increase in population, and the opening of the railroads had a positive influence. Diablos reverted to being Angelinos. In 1856, the Sisters of Charity opened the first American hospital. Soon afterward, Los Angeles saw its first library association (1858); its first charitable organization, the Ladies Sewing Society (1859); its first banks, informal (I. W. Hellman, 1865) and formal (1869); its first city lighting by gas (1867); its first volunteer fire company (1869); its first theater, the Merced (1871); its first public transportation, by horse-drawn streetcars (1873); and the opening in 1881 of the southern branch of the State Normal School (the precursor to UCLA). All these institutions made the city more livable and a great deal more genteel.

———

Throughout the chaotic and difficult period between 1850 and 1890, the growth of Los Angeles, both quantitative and qualitative, was fueled by the sale of oranges to an avid eastern market. Expelled from the Garden of Eden and told that they would have to sweat it out to make a living, humankind has turned that divine malediction into what is arguably the most insistent Judeo-Christian mandate for a utopia: find yourself a New World, go to the virgin lands, and reap personal wealth and happiness along the way.

The American Dream provides one way of understanding the tale of Southern California citriculture. But this kind of story is simplistic, inviting categorization of its protagonists as good guys and bad guys, whereas the reality is considerably more interesting. One of its most interesting features is the reformulation of the Judeo-Christian utopia under Thomas Jefferson.

In this case, the new utopia was rooted in eighteenth-century political theory. When Thomas Jefferson conceived and organized the Lewis and Clark expedition, he saw the space in the West as a blank slate upon which to write a new, more just chapter of human history. Settlers would set up a new social organization in these new territories in which they would be allotted land, with the same acreage to each family. They would live by cultivating the land. Even if they prospered, they would remain economic equals, avoiding English-style gentrification and social stratification. This was the Jeffersonian utopia, the backdrop for the colonization of the American West.

In the second half of the nineteenth century, the ideal of the Jeffersonian utopia was carried to California by small religious groups of settlers. D. M. Berry organized fifty families from Indianapolis into a group known as the Indiana Colony of California. They founded Pasadena as an offshoot of the San Gabriel Orange Grove Association. Another, similarly culturally homogeneous colony was that of the Mormon settlers who founded San Bernardino. La Verne (formerly known as Lordsburg) was likewise incorporated by members of the Church of the Brethren.

The commercial success that accrued to citrus groves as soon as the railroads had reached Los Angeles and the growers could ship crates of oranges to the East vindicated the utopian vision. This was success on a large scale. California citrus farmers at the end of the 1920s earned four times the per capita income of the average American. In 1913, the return on investment for orange growers in Southern California was about $30 million. Twenty years later, it had increased to about $145 million. This new wealth was flaunted by its proud owners: witness, for instance, some of the truly spectacular urban avenues in Pasadena. In 1895, Riverside had the highest per capita income in the entire United States. The inhabitants of such cities in the citrus belt, which extended 60 miles east of Pasadena, through the San Gabriel and San Bernardino valleys, and up to

Riverside, prided themselves on their gentility. Their cities were bursting with civic pride, and they boasted not only banks for the money they made and city halls, but also schools and colleges, libraries and art museums, luxury hotels, monumental churches, and mansions. All those satellite cities in the citrus belt, and the urban model they fleshed out, account for the multicentric character of the greater Los Angeles area.

———

What were the reasons for such prosperity in Southern California during the first third of the twentieth century, until the Depression? Some explanations combine the exploitation of minority workers hired to pick the fruit with the "great man" theory of history.

While it is quite true that non-Anglos and poor people still harvest citrus fruit, and that for a long time Chicano *braceros* (guest workers from Mexico) have provided cheap labor, all the cheaper because of illegal immigration, this was not the main source of wealth for the growers in the citrus belt. Chinese, Sikh, Japanese, Filipino, and Mexican laborers toiled in the groves. But they were not easily exploited seasonal workers, hired only for the duration of the harvest. With citrus fruit, the harvest is truly a continuous, year-round effort.

This work opportunity led to an influx of immigrants who stayed. Their arrival followed the familiar pattern: first a single man arrives. As soon as he can afford it, he sends for the rest of his family. Indeed, non-white citrus belt populations increased apace with the Anglo population during the citrus boom years 1900–1930. As a consequence, and despite the fact that citrus growers were hardly philanthropists, labor was organized early on, in the second decade of the century. Workers went on strike for higher wages and joined the IWW (Industrial Workers of the World).

Neither is the "great man" notion very helpful. Take the example of the colonists from Indiana who settled in what was to become Pasadena. One of them, D. M. Berry, was sent ahead to scout the area and find the most suitable location for their destination. He was seduced by Rancho San Pasqual, a part of the Mission San Gabriel Archángel on the shore of the San Gabriel River. When his group of followers settled there in 1873, their first effort was a financial failure. The chief merit of D. M. Berry was to trust in his initial vision and to be quick to reorganize the colony. He added a few carefully selected Westerners and, after purchasing 1,500 acres of land, incorporated the new group as the San Gabriel Orange Grove Association at the end of 1873. The land was subdivided into a hundred lots of fifteen acres each. Ten years later, two hundred families settled there (the name Pasadena, an Indian name, was adopted in 1875). What made the Pasadena colony successful, much more than the leadership qualities of D. M. Berry, was irrigation of the citrus orchards when the Wilson Ditch was dug to bring water from the nearby Arroyo Seco. We shall return to the Pasadena story later in this chapter.

––––––

Both exploited labor and visionary great men contributed to the citrus boom. But the three primary reasons for the citrus bonanza at the turn of the twentieth century were hard work, irrigation, and agricultural research. Far from being anticlimactic, the real explanation of the boom is extremely interesting—if anything, for the light it shines on the rise of the Californian economy during the twentieth century.

A citrus grove is not automatically lucrative. Earning profits year after year is hard work. What does tending an orchard entail? First, the grower has to set up a wind shield. Immediately following the purchase of land in the citrus belt between Los Angeles and Riverside, an orange grower had to erect ramparts against strong winds: high walls of trees. Eucalyptus were often selected, as they grow quickly. After only two or three years, such screens become effective, and the planting of citrus trees can proceed. The growers planted trees in rows, typically twenty-two feet apart. The young trees remained vulnerable to other aggressors, from scorching heat to jackrabbits. Hence, hired workers would wrap corn stalks around their trunks.

In order to grow, citrus trees need fertilizer, and they need water. The trees have to be watered at least once a month. Irrigation is the most important key to success, and we will come back to it.

The trees have to be sprayed regularly against various pests that may attack them. On cold winter nights, when frost lurks, they demand protection. The growers lit fires on the periphery of each orchard, which provided a cover of thick smoke against the intrusion of icy air.

And we can't forget the harvest! Harvesting citrus fruit is a year-round activity, as the fruit is best left on the trees until ready for shipment. Non-Anglo workers, typically Asians or Mexicans, would do the picking, after which horse-driven carts would take the fruit to the packers. Packing was traditionally the province of the wives of the fruit pickers. They would wash and inspect the fruits individually, selecting them for size and grading them. They would then arrange them on beds of colored paper inside crates with colorful labels. For instance, the Rialto area had, at the peak of its production, seven packing houses, four independent and three affiliated with the California Fruit Growers Exchange.

––––––

Irrigation was a nagging concern for the individual growers. The writer Susan Straight, who spent a happy, carefree childhood on a citrus farm at the base of Box Springs Mountain in Riverside, recalls "cooling [her] feet in the irrigation furrows that filled with water, making shiny silver trails through our borrowed forests. The square bricked-in water pumps released torrents into the squat, round cement towers at the head of each row, and those were our waterfalls."

Organized irrigation of the citrus groves in Southern California had to be set up as soon as plantations became widespread in the 1880s. The location of the citrus belt near the foothills of the San Gabriel mountains was ideal. Irrigation of the orange groves depended on capturing mountain streams in canals.

George Chaffey (1848–1932) made it happen. He was a Canadian engineer of many talents, a visionary as well as, arguably, a genius. His parents had retired to Southern California for their health from the area of Kingston, Ontario. When Chaffey came to visit, he was captivated by the climate and charmed by the sunny scenery and the clear air, so he stayed. He and his brother William, a horticulturist, formed a highly successful team. Among other business ventures, including some in Australia, they were responsible for the planning of a number of urban communities in the San Bernardino area.

It all began when, on Thanksgiving Day in 1881, the Chaffey brothers bought a 560-acre tract of land from Captain Joseph S. Garcia in today's Etiwanda. In 1882, the Chaffeys purchased a larger tract of more than 6,000 acres, which became the cities of Ontario and Upland. Ontario was named for the Chaffeys' original Canadian province. Etiwanda, the city built on the Garcia ranch, was named for a Lake Michigan Native American chief who excelled at community relations. Upland, Cucamonga, and Alta Loma were the other cities designed by George Chaffey.

George Chaffey's other brainchild was the Mutual Water Company. Each member of this cooperative was given equal rights to the water supply: Chaffey was intent upon preventing water wars, then all too common between newly incorporated cities in the West. He subdivided the land (with other acquisitions, the Chaffey brothers had consolidated about ten thousand acres) into ten-acre arboricultural lots, each entitled to its share of water. In order to supply the future citrus groves in Ontario, the Chaffeys purchased the Kincaid Ranch in San Antonio Canyon as a water source for this new city. Water was delivered to each lot by a cement pipe, both to prevent evaporation and to allow easy expansion of the watering network. A pipeline fed the north end of each citrus grove, and each row of trees had a standpipe at its head. Two to four gates regulated water flow from each of these standpipes.

Each grower was allotted a water supply of as much as 50 miner's inches of flow—a "head," in the technical jargon of hydraulic engineers. He would be advised of his water schedule for a 21–30-day period. Whenever the water started to flow, he had to be fully prepared, with gates adjusted and furrows at the ready. At the water company, the technician responsible for distributing the water flow on schedule to each plot was known as the *zanjero*.

George Chaffey, a versatile engineer (he was a marine engineer who had previously built steamboats in Canada), also harnessed mountain streams to turbines and generated electric power. He advertised his success with a powerful three-kilowatt white lamp over his home in Etiwanda, which could be seen many miles away, as far as Riverside. His ensuing fame was such that the City of Los Angeles hired him for its electrification. He installed the first six street lights there. Later, he became chief engineer of the Los Angeles Electric Company.

George Chaffey embodied in real life the recurrent character in the novels by his contemporary Jules Verne: the all-rational engineer, well informed of the latest scientific developments, who builds a utopian, ideal community in a desert, who makes it thrive, and who continually looks after every aspect of communal life in a benevolent, paternalistic way. Nowhere is this better epitomized than in the monumental main thoroughfare that George Chaffey gave to Ontario, California. He laid out a straight, fifteen-mile-long double boulevard, two hundred feet wide, lined with handsome tall palms, eucalyptus, grevilleas, and pepper trees. He named it, revealingly, Euclid Avenue. It was emblematic of the Euclidean rectangular grid that Chaffey superimposed on the land—in conformity with the Jeffersonian vision of a United States first surveyed, then populated and developed according to a logical, Cartesian, and geometric organization of space. Thomas Jefferson's view of American nature as a blank slate for humankind to write on was shared by George Chaffey. He marked the Southern California countryside on his blueprints, which mapped out and planned the human settlements, the colonies to come.

George Chaffey received many honors rather early on, less than twenty years after he had begun making possible the expansion of citrus cultivation in Southern California. In 1903, the federal government distinguished Ontario, holding it up as a model for American Irrigation Colonies. The same year, an international committee of landscape engineers selected Euclid Avenue, Ontario, as one of the most beautiful highways in the world. And in 1904, a miniature model of the Ontario irrigation grid was one of the attractions at the World's Fair in St. Louis.

———

We have now looked at two of the main reasons for the huge success of citriculture in Southern California: first, the well-organized hard work of the growers and their workers, and second, an irrigation scheme that was technically efficient and democratically run. The third, equally important factor was science. From very early on, state authorities recognized the importance of arboricultural research.

By the early years of the twentieth century, fruit growers could draw on the expertise of agricultural research stations established by the California Commission of Horticulture and the University of California system. As we have seen, a Citrus Experiment Station was set up in Riverside in 1913. Such laboratories studied geography in order to select the best locations for each citrus variety. They compiled data on the hardiness of trees relative to soil type and climate. Arboricultural research went on to optimize fruit size, ease of peeling, aspect (color and texture), sugar-to-acid ratio, and hence taste. It designed desirable new qualities, such as seedlessness, through hybridization.

The citrus varieties planted at the outset, in the 1880s, had been Mediterranean sweet oranges, St. Michael's oranges, lemons, and the *umbigo* orange from Bahia, Brazil, which came to be known (through the agency of the federal government) as the Washington navel. At the peak of Southern Californian production, in the late 1920s and early 1930s, the three basic varieties cultivated in the orchards from Pasadena to Riverside were grapefruit, Valencia oranges, and Washington navel oranges.

Among successful innovative varieties, the most noteworthy were, among oranges, the Trovita, a non-navel seedling raised in 1914–1915 at the Citrus Experiment Station and released in 1935; among lemons, the Lisbon, which was nurtured by H. B. Frost at the Citrus Experiment Station in Riverside in 1917 and released about 1950; and among tangerines, the clementine, an accidental hybrid. Introduced into the United States in 1909, it was brought to California from Florida in 1914 by H. S. Fawcett of the Citrus Research Center, Riverside.

Citrus varieties sometimes hybridize accidentally in the wild. Some of the most fascinating examples come from crosses between the sweet orange (*Citrus sinensis*) and the common mandarin (*Citrus reticulata*). They are known as "tangors," a composite word from *tang*erine and *or*ange. Thus the name Ortanique, a name I love. It rhymes with Martinique, which, to me, enhances the taste of the Caribbean that the fruit carries from its origins in Jamaica. Its discovery and its subsequent history are the stuff of tall tales. A man named Swaby from Manchester, Jamaica, found a half-dozen or so specimens of an unusual fruit in the Christiana market and planted the seeds. Of the ensuing seedlings, two bore fruits, which were exhibited at an agricultural show at the turn of the twentieth century. C. P. Jackson, of Mandeville, Jamaica, bought two of them and planted 130 seeds. Jackson then selected the least thorny and least seedy specimens. He coined the name Ortanique from *or*ange, *tang*erine and un*ique*. In California, the Citrus Growers Association took charge and started exporting the fruit in 1944. It is a fruit in high demand, and it brings a premium price.

What nature achieves by chance, man attempts by choice. Walter T. Swingle, at the USDA nursery in the Little River district of northeastern Miami, also succeeded in hybridizing *C. sinensis* and *C. reticulata*. Around 1913, he sent the ensuing tree for trial to R. D. Hoyt in Safety Harbor, Florida. Hoyt passed on budwood to his nephew, Charles Murcott Smith, who propagated it in about 1922. This plant became reasonably successful, so that several nurseries started commercializing it in 1928 under the name Honey Murcott—or just Murcott. The fruit, which handles and stores well, has a dozen segments and an abundance of red-orange juice. It sells very well, but only as a fresh fruit, since the flavor of the juice is altered by processing.

Tangelos are another kind of hybrid, this time between a mandarin (*Citrus reticulata*) and a grapefruit (*Citrus paradisi*). Some tangelos are accidents of nature; the best known of these is the Ugli. Other tangelos are man-made; among these are hybrids called *swinglets* (a tribute to Walter T. Swingle).

As early as the 1890s, Walter T. Swingle and Herbert J. Webber independently developed a hybrid of the Bowen grapefruit and the Dancy tangerine. This cross became known as the Minneola. Its tangerine ancestry is marked by the vibrant orange of the flesh and a deep orange-colored peel that is easy to remove. The grapefruit endows it with fewer seeds (seven to twelve small ones) and a medium-large size, about 3 1/4" × 3". The taste is described as a sweet citrusy flavor leaning toward tangerine, with honey and grapefruit undertones. A smidgen of tang balances the sweetness. Well done, Drs. Swingle and Webber!

In 1911, Swingle crossed the same two varieties again, this time a female Duncan (or Bowen) grapefruit with a male Dancy tangerine. The offspring became known as the Orlando, and it has become a good commercial fruit in Florida. Its pollination is best achieved by honeybees, for a better yield. The medium-sized fruit (3″ × 2 3/4″) has a deep orange skin, slightly rough and not loose; twelve to fourteen segments; and ten to twelve seeds. It is very juicy and sweet.

———

All these new hybrids, and all this activity, notwithstanding, expulsion from Paradise was inevitable. The Southern California citrus belt, running from Pasadena to Riverside, is now only a receding memory. What happened? A visit to Caltech is instructive, as one might expect.

The California Institute of Technology—Caltech—is a world-renowned beacon of excellence. As an institution of higher education, it is remarkable, whether for its per student endowment (it's known as "The Most Expensive University in the World"), its faculty-to-student ratio (the student body is tiny), or the average test scores of its entering classes. It is located in Pasadena, which today is a plush small city about fifteen miles to the east of Los Angeles International Airport and about five miles from downtown Los Angeles. The Caltech campus, of a size in keeping with the excellence of the institution (it takes five to ten leisurely minutes to walk across it), is well tended, with some lovely buildings.

Furthermore, a visitor to the campus notices anonymous historical markers—orange trees. The turf beneath them is littered with fallen oranges. The visitor might also notice, with a chuckle, discarded and burst orange hulks far from the trees—obviously the remnants of late-night mock battles between students. These young people carry huge expectations on their shoulders, from family, staff, and even outside businesses. Throwing a few oranges must be a pretty good relief from the ensuing stress.

The trees are a reminder that Pasadena originated as a community of citrus growers and packers, and that Caltech itself succeeded an earlier university, Throop Polytechnic Institute, founded on the new wealth made from shipping the golden fruit of the California sun to the eastern United States. Nowadays, the citrus groves are long gone. But the city of Pasadena endures—and not only because it boasts Caltech.

The prosperity of the Pasadena area originated in the 1880s. Pasadena was incorporated as a city in 1886. The two hundred families of 1883 grew to nearly 10,000 inhabitants by the turn of the twentieth century, and three times as many only ten years later. Throop University, the seed for the Polytechnic Institute (and later Caltech) was founded in 1891.

Wealth attracts wealth. Pasadena at the turn of the century was a fashionable address for tourists, and better yet for new settlers. Developers built luxury hotels. South Orange Grove Avenue became aptly known as "Millionaires' Row." Well-known architects opened offices in Pasadena; in particular, the firm of Greene and Greene, two brothers, built some magnificent, style-setting mansions there. All this prosperity came from the orange groves.

Then why did the orange groves disappear—to such an extent that, a century later, the orange trees on the Caltech campus serve as only a nostalgic reminder? Pasadena—like many other communities in the Southern California citrus belt—became a victim of its own success.

The brightly colored and extremely attractive citrus crate labels, emblematic of advertising by the citrus belt, did too good a job: tourists flocked to these idyllic places, which seemed to guarantee gorgeous weather, crystal-clear views, beautiful mountains as a backdrop, opportunities for horseback riding and strolls in the countryside, and, permeating the air, the addictive and ever so delicate scent of orange blossoms.

The new twentieth-century Los Angeles, with its booming satellite urban centers, started to expand in a big way, and most of the expansion was in the citrus belt. As the decades rolled on, the tourists-into-settlers-into-retirees pattern seeded newer communities. At the same time the citrus groves and the attendant budding agribusiness were becoming established, other industries arrived, first and foremost the movie and entertainment industry.

All these factors meshed so that when the Depression hit the citrus industry in Southern California (due to the collapse of its main markets in the East), the agricultural lands were sold at rock-bottom prices to investors. World War II was a period of renewed economic boom in Southern California due to the activity of the shipyards and the aerospace industry (companies such as Douglas, Lockheed, and Hughes). By the end of World War II, there came another massive influx of people. California's economy became one of the most prosperous in the world, outstripping those of many nations, including industrial ones. By 1950 or so, most of the citrus groves were gone from Southern California.

The orange blossom–scented air is now to be found elsewhere in California, in the Central Valley. How did this happen?

Urbanization in the greater Los Angeles area chased the citrus growers away, and they relocated. Their needs, remember, were irrigation, cheap labor (to some extent), agricultural research, and the two S words, sun and soil. The Central Valley of California provided all of these, with the help of the federal government.

The last two factors, the S words, are just about ideal in the Central Valley. The valley is a trough about 50–80 miles across, extending for more than 400 miles from Redding to the Tehachapi Mountains, along a northwest-to-southeast axis, in between the coastal range and the Sierras. Seemingly flat, it is filled with rich, porous alluvial deposits from rivers, mostly the Sacramento River in the north and the San Joaquin River in the south. The sun is lavished on the valley in scorching quantity during the summer, when temperatures above 100°F are common. The soil-sun combination is a winner, ideal for most crops. The citrus growers made their move there in the aftermath of World War II, and they have settled, en masse, in a narrow thermal belt south of Fresno, close to the mountains on the east side. Three counties in the Central Valley, Fresno, Tulare, and Kern, rank first, second, and third among the top ten agricultural counties in the United States. The predominant towns of the new citrus belt are Woodlake, Exeter, and Porterville.

Before this could happen, however, one problem had to be solved: the climate of the Central Valley is desertlike. The annual rainfall is less than five inches in southwestern Kern County, and only about ten inches in Fresno. This is where the federal government stepped in. As early as 1924, the Army Corps of Engineers began projects in the Central Valley for both navigation and flood control. As an answer to the Depression, Franklin D. Roosevelt started numerous public works projects. One of the most monumental, together with the TVA in Tennessee, was the Central Valley Project (CVP), which began in 1937. The idea for the CVP could not have been simpler, but it was nonetheless audacious: it took the water from the north and gave it to the south to make the crops thrive. In other words, it diverted surplus water from the Sacramento River and its tributaries into the water-deficient areas of the San Joaquin Valley.

The availability of water in the south of the Central Valley had been an issue for generations. The residents of the town of Lindsay in the 1920s, starved for water, sank 39 wells near the Kaweah River, near Visalia. This dried up the water supply of Visalians downstream on the Kaweah. The Visalians filed a lawsuit, which went all the way to the Supreme Court, and this particular water war raged for two decades.

The CVP put an end to these water wars. The Contra Costa Canal started delivering water to the region in 1940. Work on the Shasta Dam, the central feature of the CVP, started in 1937. Even before its completion in 1945, water was stored behind the Shasta Dam.

The main water supply to the present citrus belt was put in place at the end of World War II. Northeast of Fresno, the Millerton Reservoir, behind the Friant Dam, is fed by the San Joaquin River. Two mainline canals were built from the reservoir, the Madera Canal heading north and the Friant-Kern Canal heading south. Ninety-five percent of the water trapped at Millerton goes to agriculture. The Friant-Kern Canal, operational since 1944, runs for more than 150 miles to the Kern River, near Bakersfield. This irrigation makes possible the $2.1 billion per year agribusiness of citrus farming.

The Central Valley is also in an ideal location for shipping its produce. Several highways run along it. U.S. 99 is a fast thoroughfare with immense straight sections, whose monotony is enlivened by the colorful oleanders along the center strip. Parallel to it runs Interstate 5, also easily accessible from the citrus belt for southbound transportation to Greater Los Angeles.

Scientific research is provided by agricultural stations set up by the State of California and on University of California campuses, primarily at UC Davis, arguably the world's leading university for agricultural research. Davis is easily accessible from any point in the Central Valley, while smaller campuses—such as Fresno State University—provide education for the population of the eight-county San Joaquin Valley.

The citrus growers brought with them their social organization— basically a legacy from the plantation system in the American South, with ethnically distinct manual laborers, but without the stigma of slavery. The population of the Central Valley in the 1940s was 20 to 30 percent Hispanic. The newer arrivals from Mexico, as often as not legal, still perform the more menial chores of pruning, planting, and picking. Nowadays, due to the bold action of the charismatic Cesar Chavez, many have become organized under the banner of the United Farm Workers (UFW). And, as elsewhere in the Central Valley, Anglo families (often descendants of the Okies immortalized by John Steinbeck's *Grapes of Wrath*) are now at the top of the totem pole.

The new citrus belt has not yet become a tourist attraction. It is said that you smell the city of Lindsay, in the heart of the new citrus belt, before you see it. April is the month when the fragrance from the orange blossoms is at its peak. The heavenly scent, which has deserted Pasadena, Alta Loma, and Riverside, now hangs over Lindsay, Porterville, and Ivanhoe.

The crate labels of the 1880–1950 period, advertising the old citrus belt in Southern California, have become museum pieces. The old belt's production centers are now history. Pasadena is a modern Rome for the historian of citrus production—and for the art historian, too: the Norton Simon Museum houses one of the most beautiful paintings ever made, Zurbarán's *Lemons, Oranges, and a Rose.*

As mentioned at the beginning of this chapter, the California success story had its counterpart in Florida—and not only there, but elsewhere. I close this chapter by recounting how, in a part of the world very distant from its origins, citrus fruit was deliberately chosen as a cornucopia, for providing a livelihood—again by Anglos of sorts, the New Zealanders.

———

To New Zealanders, the northernmost part of North Island is the barren Great North. It takes three hours to drive due north from Auckland to Kerikeri, the little town where most of New Zealand's citrus groves are located. Seen from the air, it is a mosaic of small, carefully manicured enclosures (about twenty acres each). On the ground, the visitor discovers that the plantations are indeed protected from the winds by tall furry hedges of trees, predominantly rimu pines, bamboos, eucalyptus, and *Hakea saligna.*

In 1926 or so, at a time when the strife in China that would become the civil war between the Kuomintang and the Communists was beginning, E. S. Little, a British businessman in China, saw the writing on the wall. He decided to emigrate to faraway New Zealand, where he was instrumental in developing the North Auckland Land Development Company. He thought the volcanic soils in the Kerikeri area would be ideal for citrus growing. His company surveyed the land, divided it into lots, and brought in citrus plants from China. Other British exiles from China, many from the Shanghai area, followed his example. A man named Bus Emanuel, for instance, settled on one of the orchard lots in 1927. By the outset of World War II in 1939, the Emanuel lemon orchards were the largest in the Southern Hemisphere.

Citriculture brought prosperity to the area, which even today contrasts with the all-too-obvious poverty of the New Zealand North. Kerikeri is a wealthy small town of about 3,000 people, not counting the tourists flocking to the adjoining spectacular Bay of Islands area in the summer. The good times lasted until 1984, when a Labour government (headed by David Lange as prime minister with Roger Thomas as minister of finance), influenced by the market liberalization policies of Ronald Reagan and Margaret Thatcher, put a brutal end to New Zealand autarky.

Until then, the Kerikeri citrus growers enjoyed an absolute monopoly on the production and sale of citrus fruit, which made them quite wealthy. Up to that point, the New Zealand citrus story is practically a model of the kind of success that was enjoyed—with perhaps more practical difficulties—by Florida, Brazil, and especially California.

SIX

Making Lemonade out of Lemons

Ajo, cebolla, y limón, y déjate de inyección.
(Garlic, onion, and lemon, and you can drop the injections [i.e., their ingestion will keep you healthy].)
SPANISH PROVERB

A diminutive boy stood in a New Orleans emergency room. Jim (not his real name) had a distinctive appearance. At the age of six, while he stood 3′10″, he weighed only 37 lbs. The family was familiar with physicians. Jim had a four-year history of autism on top of his delayed growth and physical development.

Jim's parents brought him to the emergency room because he was suffering from pain in his hips. He had been limping for the last six weeks. The day before, Jim had suffered so much pain that he was unable to walk.

A blood test showed normal hemoglobin and cell counts. A test for rheumatoid factor was negative, as were other tests for arthritis. X-ray pictures of Jim's legs showed diffuse osteopenia—bone tissue depletion. MRI scans of Jim's pelvis and hips did not show anything abnormal. Even though Jim's white blood cell count was at worst borderline, the physicians at Children's Hospital in New Orleans suspected acute leukemia at first. A biopsy and a bone marrow smear were negative, however, ruling out leukemia.

Finally, they came up with a diagnosis: the boy had scurvy. This disease, caused by a deficiency of vitamin C, has become so rare in developed countries that physicians, as a rule, almost never come across a case.

Indeed, Jim's diet during the previous year had been unbalanced, consisting only of cookies, yogurt, whole milk, biscuits, and water. He never ate fruit, vegetables, meat, or fish. The cure confirmed the diagnosis. A high dose of vitamin C (200 mg daily, orally) for ten days and a more balanced diet were enough to reverse all of Jim's symptoms, which included swollen and bleeding gums—another telltale sign.

The main symptoms of scurvy are general fatigue; extreme sensitivity and pain in the whole body; purple skin spots, especially from the waist down; and swollen gums and loose teeth. Without treatment, the weakness gradually worsens, and death ensues.

Scurvy: the word alone no longer frightens. The disease has become a mere curiosity, relegated to the history books. Humankind is so oblivious of the torments once caused by scurvy that they have committed this scourge to history.

That a little boy in Louisiana fell victim to scurvy is now only a footnote in the daily barrage of information. News agencies and the world media may feature such a story for a day, and then they move on. But for centuries, scurvy was dreaded. Sailors the world over deemed it an inevitability.

Witness George Anson's circumnavigation of the globe in 1740. Anson (1697–1762), one of Britain's most remarkable seamen, was entrusted with a flotilla of six ships and a crew of 2,000 sailors to travel around the world. His strategic goal was to cripple Spain economically; he was looking for ways to jeopardize her trade in the Pacific.

Seven months after leaving Portsmouth, Anson's expedition reached Cape Horn, and the crews started showing symptoms of the disease: "large discolored spots over the whole surface of the body, swelled legs, putrid gums...extraordinary lassitude of the whole body... ulcers of the worst kind, attended with rotten bones, and such a luxuriancy of fungous flesh as yielded to no remedy," as described by the expedition's chaplain, Richard Walter, who published a memoir of the journey, *Voyage Round the World*.

By the time Anson had rounded Cape Horn, his expedition was already a disaster. Two of the ships had to turn back; a third was wrecked. On the three remaining ships, 626 of the 961 men were dead—from scurvy. When Anson returned home in 1744 with a single ship, fewer than 200 men out of the original 2,000 had safely completed the voyage.

This sad tale was no fluke or isolated incident. It was typical of the predicament of many Royal Navy vessels on long voyages. Did scurvy affect other countries' navies? The Portuguese were somewhat less beset by the scourge on their naval expeditions to the East.

Because of the prevailing winds, the best sailing route from Portugal to India crosses the South Atlantic, following the coast of Brazil prior to veering eastward to South Africa and around the Cape of Good Hope. Indeed, this was how Pedro Alvares Cabral came upon Brazil "accidentally" in 1500, on his way to India. A letter that Father Manuel da Nóbrega wrote home in 1549 suggests that at that time, the coast of Brazil was heavily planted with lemon, orange, and citron trees.

Both Arab and Portuguese navigators had some knowledge of the use of citrus fruit against scurvy. This conclusion is based on three pieces of information. The first is a note made by Alvaro Velho in his journal for January 7, 1499, during the first trip to India led by Vasco da Gama, when the ship stopped at Melinde, on the coast of present-day Kenya, on the return leg of the voyage:

On Monday, we stopped in front of Melinde, where right away the king sent us a long bark, which brought a crowd, and he sent us sheep; and our captain sent back to shore a man, together with our visitors, to bring back the next day oranges, which were much in demand by the sick people [from scurvy] whom we carried; which he did promptly, together with much other fruit.

The second is testimony by Thome Lopes upon Vasco da Gama's return from his second voyage to India. In his report of August 21, 1502, he described a purchase of oranges and other foodstuffs from Melinde, which had a marvelous effect on the health of the sailors, bringing relief from the swollen gums and other symptoms of scurvy.

The third is the possibility that the Jesuits—who were influential in the dissemination of plants from both India and Portugal in newly discovered Brazil—recommended planting citrus trees on the Brazilian coast, with a view to alleviating symptoms of scurvy among Portuguese sailors.

This precious information was not available only to the Portuguese. During the sixteenth century, Dutch sailors departed the Iberian Peninsula with lemons and oranges as part of their cargo. In 1605, Sir James Lancaster, a captain in the British Navy, gave two daily spoonfuls of lemon juice to the sailors on his flagship, the *Dragon*, but not to those on the other three ships in his squadron. There were no deaths on the *Dragon*, but a 45 percent mortality rate on the other vessels.

By the eighteenth century, such wisdom had inexplicably vanished. The consequences were disastrous, to the British Navy in particular. One reason for its disappearance may have been distrust of folk remedies, in this case the recourse to fresh fruits and vegetables. Until the American and French revolutions, the pragmatic knowledge of artisans and manual workers was deemed inferior to formal knowledge from books, especially with respect to medicine. There were some exceptions, such as Bernard Palissy (ca. 1510–1591), a French potter of genius and a pioneer in experimental science. But the prejudice endured until Diderot and D'Alembert engaged in a deliberate attempt at rehabilitation with the *Encyclopédie*. Looking down on manual work and folk knowledge originated with writers in Greek antiquity, Plato in particular. Even if pragmatic knowledge about the use of citrus juice existed, it was considered unworthy of the attention of educated people of the ruling class, and they chose the diet for sailors on the ships of the Royal Navy.

Another reason might have been that the Portuguese kept the use of citrus fruit to prevent and/or cure scurvy a secret. It was strategic information, given the foremost economic importance of the Spice Route by sea from Southeast Asia and India to Europe, which took months to travel. Such a vital piece of information could well have been entrusted only to the chief of an exploratory voyage, such as Vasco da Gama.

History sometimes stutters, forgetful of earlier achievements. But history often recalls them, sometimes centuries afterward. Linear perspective, for example, known during antiquity, was rediscovered by Alberti during the Renaissance, just as citrus fruit was rediscovered as a preventive and cure for scurvy.

In the eighteenth century, a Scottish physician for the British Navy, James Lind (1716–1794), proved that citrus fruit protected against scurvy. But where did he get the idea to try lemon juice in his historic experiment? Clearly, this was not fortuitous. He built on what was common knowledge among some people.

In 1747, Lind performed an admirable experiment—arguably the first clinical trial in history. While on a cruise in the Bay of Biscay, he tested a citrus fruit cure on sailors sick from scurvy. This field trial compared alternative treatments, mostly acidic substances recommended by various authorities. Two of the men received two oranges and one lemon daily. Another two patients were given a daily quart of cider. Two swallowed two spoonfuls of vinegar, thrice daily. Two got, again three times per day, twenty-five drops of elixir of vitriol (sulfuric acid!). And another two were administered, also three times daily, an elixir made of cream of tartar, barley water, horseradish, mustard seeds, garlic, Peruvian balsam, and tamarinds.

The two patients on citrus improved spectacularly. After six days, one of them even resumed active duty. James Lind went on to write and publish, in 1753, *A Treatise of the Scurvy*. Is this 454-page monograph to scurvy what *Moby Dick* is to whale hunting? Not quite: it is predominantly an altruistic pamphlet denouncing the dreadful conditions—the dank and unhygienic spaces—in which sailors lived. Lind devoted only two pages to the citrus treatment.

The bureaucracy of the admiralty was unconvinced, and remained inactive. Not until 1795 would British seamen start receiving a regular dose of lime juice—which earned them abroad the derisive name "limeys." During the nineteenth century, the nickname would extend to Britons in general.

Lime juice proved quite effective: The Royal Naval Hospital at Portsmouth housed 1,457 cases of scurvy in 1780, but only 2 in 1806. From the beginning of the nineteenth century, scurvy started to become a pathology of the past.

Nevertheless, the empirical finding that citrus fruit prevents scurvy was not sufficient. It remained to understand why. What was the active factor in lime juice, lemon juice, or orange juice? This piece of the puzzle turned out to be most elusive. In spite of the efforts of many scientists, that factor escaped isolation, purification, and characterization until the twentieth century.

———

The ingredient in citrus fruit that prevents scurvy was as difficult to identify as the proverbial needle in the haystack. Citrus fruit is a treasure trove of chemicals that are highly useful to humankind. There are dozens of such chemicals, and I shall mention only a handful here.

Hesperidin, easily extracted from orange peel (college undergraduates perform this extraction routinely in many organic chemistry laboratories), is easily converted into a compound known as neohesperidin dihydrochalcone (NHDC). NHDC is a potent, low-calorie sweetener, up to 2,000 times sweeter than sucrose (ordinary sugar) per unit weight. Narangin, also present in orange peel, can also be converted into NHDC.

Limonene, derived from the lemon, as the name indicates, and present in the peel of citrus fruits in general (it constitutes about 98 percent of the weight of the essential oil from orange peel), has many applications, ranging from commercial floor waxes to perfumes, and it is the primary material for making menthol, the main source of mint flavor. I surmise that the reason for the presence of this chemical in citrus peel is that it is a natural insecticide—and there are other chemicals in citrus peel that are similarly effective in repelling insects. Limonene will keep most external pests, such as lice, fleas, and ticks, away from pets.

Citric acid, whose name also emphasizes its presence in citrus fruit (it is the major acidic constituent in the juice), is one of the most important biochemicals on earth. The so-called citric acid cycle, first described by Hans Adolf Krebs (1900–1981), gained him the Nobel Prize in 1953, and is a cornerstone of biochemistry.

Lycopene, a molecule of the carotenoid family, coexists in citrus peel with chlorophyll and is responsible, for instance, for the yellow-pink color of grapefruit and the red color of tomato juice. It is vaunted for reducing the risk of atherosclerosis and coronary heart disease. Like other carotenoids, it has also been promoted as a preventive and cure for epithelial cancers, such as those of the lung, bladder, cervix, and skin.

Any citrus fruit, then, contains a most impressive pharmacopoeia. But none of the molecules just mentioned acts to prevent scurvy. Isolation of the responsible factor took many years. Progress in understanding the effectiveness of citrus fruits, and other fresh fruits and vegetables, against scurvy came very slowly. In 1905, a Dutch professor, Cornelius Adrianus Pekelharing, writing of his experiments on mice, noted that "there is an unknown substance contained in the milk, which even when the intake is extremely small, is of the utmost importance for nutrition." He showed that even in the seeming midst of plenty (of fats, proteins, and carbohydrates), if this "unknown substance was missing, the mice would die." Unfortunately, his report was published only in Dutch and had a limited audience.

In 1927, the Hungarian Albert Szent-Györgyi (1893–1986), then a professor at the University of Szeged, isolated a substance that he called his "Groningen reducing agent" in a crystalline form from oranges, lemons, cabbages, and adrenal glands. Tongue in cheek, he even proposed a name for the new substance. Sugar molecules have names ending in -ose, such as gluc*ose* or fruct*ose*. Thus, Szent-Györgyi proposed to name his crystalline sample "ignose"—thus pointing to its apparent relationship to sugars while underscoring his *igno*rance of its true nature. But Arthur Harden, the editor of the *Biochemical Journal* at the time, "did not like jokes and reprimanded me." A second suggestion, "goodnose," was judged to be equally unacceptable. Szent-Györgyi finally agreed to accept Harden's somewhat more prosaic suggestion, "hexuronic acid," "since this molecule had six carbon atoms and was acidic."

Which brings us to the early 1930s. To set the stage, physiological chemistry (biochemistry dealing with normal and pathological conditions) was then practiced predominantly in Europe and in the United States. However, in paternalistic Europe, any topic that dealt with nutrition was looked down upon. It smelled of the kitchen and was deemed too feminine for serious scientific work. Hence, ambitious male scientists such as Szent-Györgyi spurned any work in such areas.

By contrast, Americans were eager to investigate and identify the anti-scurvy molecule, which came to be known as vitamin C. They sensed that a Nobel Prize awaited its discoverer. Such an accomplishment would be a first: no native-born American had yet won the Nobel Prize in Physiology or Medicine for a discovery benefiting humankind. So there was frantic competition to establish the structure of this particular substance. However, the substance was hidden among the many different sugar molecules present in citrus juice, and it remained elusive.

To compound the difficulty, this particular chemical is a vitamin for only three species of land mammals besides humans. A *vitamin* is a chemical that is essential to the body and its well-being, which nevertheless is not made by the body and thus has to be ingested in food. The only laboratory animal also needing vitamin C in its diet is the guinea pig. And guinea pigs are unwieldy laboratory animals; they are more expensive than mice, and they are voracious eaters.

One of the ambitious Americans was Charles Glen King (1896–1988), at the University of Pittsburgh. His was the logical approach: first isolate and purify the vitamin, and then determine its structure—the concatenation of atoms making up the molecule.

In the early 1930s, one of King's coworkers, a young American by the name of Joe Svirbely, went to Hungary with a postdoctoral fellowship to study with Professor Szent-Györgyi. Upon arrival, he found that Szent-Györgyi had, altogether fortuitously, isolated and crystallized a sugarlike molecule that he had named hexuronic acid. No wonder that, early in the winter of 1931, shortly after Svirbely had arrived in Szeged, the young American, trained in the race for vitamin C, told his Hungarian supervisor: "I now know what your crystals are. They're pure vitamin C." Ensuing tests proved him right.

At about that time, King wrote to Svirbely, asking what he was up to. Svirbely, prior to sharing any information with his former mentor, sought Szent-Györgyi's permission. And the Hungarian scientist, both enthusiastic and trusting, told him to go ahead, sparing no detail, and to announce to King their joint success in identifying vitamin C with Szent-Györgyi's hexuronic acid.

King moved quickly. On April 1, 1932, he published in *Science* a letter announcing his discovery of vitamin C and its identity with hexuronic acid. He made no mention of Svirbely, Szent-Györgyi, or their work at the University of Szeged. The letter in *Science* made quite a splash, including a front-page article in the *New York Times*.

Back in Hungary, Szent-Györgyi was incensed: this approached outright theft. He had been scooped in what was obviously one of the major discoveries of the century. He felt all the more betrayed because the information had reached King due to his own insistence on full disclosure. Szent-Györgyi and Svirbely rushed into print themselves, and their letter appeared in *Nature* on April 16, 1932. Hexuronic acid was renamed ascorbic acid in 1933 by Szent-Györgyi and W. N. Haworth (1883–1950).

The committee that became involved in what resembled a paternity suit over vitamin C was the Nobel Committee in Stockholm. As a rule, it draws on multiple sources of information, obtaining advice from universities in various countries and from prominent individual scientists. It also uses previous winners and senior statesmen of science as consultants—in the latter case, probably including Edward Calvin Kendall (1886–1972).

King did not improve his chances for a Nobel Prize by his 1933 application for a patent on vitamin C and its manufacture from lemons, which failed. After a long fight, King and his coworker W. A. Waugh were finally granted U.S. patent no. 2,233,417, in 1941, for their isolation of vitamin C from lemon juice.

In the meantime, the Nobel Committee had awarded its 1937 prize to Szent-Györgyi. King was not included. The discovery of vitamin C was deemed to be an achievement of the first rank in the 1930s. Not only did the 1937 Nobel Prize in Physiology or Medicine recognize its identification, but in the same year, Haworth and Paul Karrer (1889–1971) received the Nobel Prize in Chemistry for determining the structure of vitamin C.

For some reason, vitamin C continued to be prone to similar disputes and controversies. In 1933, the molecule was synthesized simultaneously, but independently, by Tadeus Reichstein in Switzerland and by Haworth and his colleagues in Birmingham, England, with both groups using essentially the same method. The Swiss group published their results just ahead of the Birmingham workers.

———

So, scurvy is ascribable to a lack of vitamin C, or ascorbic acid, and a diet poor in fresh fruits and vegetables causes it. In past centuries, sailors were a group at risk. As a rule, they were at sea for months without supplies of vitamin C. The body can store ascorbic acid for only about three months, and then scurvy strikes.

A few residual questions remain. Why do human beings need ascorbic acid in their diet, instead of producing it themselves? In other words, how did we lose the ability to synthesize vitamin C, which most animals have? And, on the other hand, how could human populations anywhere on the planet survive without access to the usual sources of vitamin C, fresh fruit and vegetables? And why is it that, among the dozens of different chemicals present in citrus fruit, conventional wisdom holds vitamin C to be of first and foremost, if not of singular, importance?

The answer to the first question we do not know. We can only speculate that, if indeed humankind had an African origin, for which there is evidence aplenty, its nutrition drew heavily on fruits rich in vitamin C. This is true of present-day African hunter-gatherer populations; for instance, the !Kung subsistence diet includes thirty species of berries and fruit.

The answer to the second question is the ingeniousness of hunter-gatherers. Take the Inuit, inhabitants of the Arctic from northwestern Alaska to Greenland, as well as of coastal regions in eastern Siberia, the Bering Strait islands, and more southerly regions of Alaska. The traditional diet of the Inuit includes meat eaten raw or undercooked, which is a source of vitamin C. Thus the Inuit shielded themselves from scurvy; they started contracting it when "civilization" reached them in the form of missionaries, when they left off the practice of eating practically raw meat.

The answer to the third question, why vitamin C is considered so important, evokes horror stories about whole crews of ships decimated by scurvy during early modern times. This is what put vitamin C at the forefront of public consciousness, as exemplified by the 1937 Nobel Prizes. It also helps to explain the popular echo when another Nobel laureate, Linus Pauling (1901–1994), promoted vitamin C as an innocuous cure-all. He pushed for the taking of vitamin C in high doses as a preventive and cure for the common cold as well as a cancer fighter. Both claims turned out to be unjustified by the usual human clinical trials, in which groups of patients received either the drug or a placebo.

Today, it is common knowledge that we need to absorb vitamin C by including fresh fruits and vegetables in our diet. It is also generally known that citrus fruit is rich in the vitamin. A five-ounce navel orange provides about 75 mg of vitamin C.

I shall turn here to a personal memory. An integral part of a climb in the Alps, which I often undertook when I was young, is the food and drink you carry in your backpack. As a youngster, your mother prepares it for you. Later on, the innkeeper, whether in a hut, a hostel, or a regular hotel, takes on this role and supplies your provisions for the day. While the composition of the food varies, a climber's pack generally contains bread and cheese, sandwiches of various kinds, dried fruit, and some chocolate or candy for quick caloric intake. Often, whatever the season, an orange or two are part of the assortment.

I have a vivid recollection of making such a trip on skis (outfitted with sealskin on the underside for a better grip on the snow) at Easter time. My friend Alain Schilling and I left the Val d'Isère resort, about 6,000 feet in elevation, for a nearby mountain known as La Grande Motte, a summit about 12,000 feet high. At that time no lifts had been built, so you had to make the climb yourself. While not technically difficult, it was a long climb. We made our way in the glare of the sun reflecting on the snowy slopes. I was drenched in sweat.

Alain and I reached the summit early in the afternoon, having left the village after breakfast, around 7:00 a.m. By that time, our water bottles had long been empty. I remember the exquisite taste of our oranges when we opened our backpacks on the summit—we truly felt on top of the world. The fruit was, of course, very cold, much colder than usual because of the below-freezing temperatures of the morning. The orange segments tasted like sherbet in our parched mouths, all the more delicious because of our thirst and our need for calories after the long, strenuous climb.

———

There are many recipes for orange sherbet. The following is among the easiest to implement: it uses only orange juice and evaporated milk, and does not require an ice cream maker.

ORANGE SHERBET

3/4 cup orange juice
3/4 cup granulated white sugar
1 cup cold milk
1 5 oz can evaporated milk, refrigerated

- In a bowl, mix the sugar with the juice, stirring until all the sugar is dissolved.

- Stir in the milk gradually, until it is also fully incorporated and the mix is homogeneous.
- Pour into a shallow dish and store in the freezer until firm.
- Remove from refrigerator, break into small chunks, and beat with an electric mixer until the preparation is smooth.
- In a separate bowl, whip the evaporated milk until it is stiff (which pumps air into the mixture to form a firm and stable foam).
- Fold into the previously frozen mixture. Return the mixture to the freezer.

To prepare an orange container to hold the sherbet:
- Slice off the top of an orange, remove, and set aside.
- Through the opening, scrape out the entire contents of the orange.
- Fill the orange with sherbet, replace the top, and freeze again before serving.

In France, this dessert is called *orange givrée.*

———

Citrus also enriches many drinks, as can be seen in the French vocabulary for the various renditions of lemons combined with different kinds of water. In a French café, whenever you order a glass of Perrier sparkling water and request that a slice of lemon be added, the waiter will repeat it loudly as "Perrier citron tranche." *Citron*, a word whose nasal ending may be a little rough on the American tongue, is French for lemon. And *tranche*, with another nasal vowel, means "slice." The result is wonderful, good reason for the detailed accuracy of your order.

Another variant, had you so wished, would have been a *citron pressé*: a tall glass filled with lemon juice would be brought to you, together with a carafe of iced water; you would add sugar to your taste, and stir it with a long spoon. If you had asked for Perrier instead of regular water, you would use a different name.

And there is yet a third type of lemon-flavored Perrier. In this drink, instead of freshly squeezed lemon juice, a dash of lemon syrup would be added; the resulting concoction tastes to me like an industrial preparation, sickeningly sweet and with a synthetic lemon aroma. This would be a *Perrier sirop de citron*, often referred to as a *Perrier citron* for short.

But back to the *Perrier citron tranche*. The glass is a miniature science demonstration that bears watching. If a fresh, unopened bottle of Perrier is used, opening it reduces the pressure on the contents, and some dissolved carbon dioxide is released. Then you will see a fascinating process, as the dissolved molecules congregate into tiny bubbles of gas, which then shoot up rapidly and burst at the surface of the water as the glass is filled. Watch these bubbles shooting up. Their size increases, since the pressure and the volume of any gas are inversely related. The very same phenomenon inflates a balloon, causing it to rise. Each carbon dioxide bubble in your Perrier is like a tiny balloon. Moreover, the trajectory of any such "balloonicule" is straight up, a handsome demonstration of gravity at work: this knowledge may be useful if you are ever caught in an avalanche.

But examine the slice of lemon floating on your drink. Its underside is coated with bubbles of carbon dioxide, large and small. Why? Very simply, because the lemon slice caps the vertical column of dissolved carbon dioxide. It traps bubbles released from below. Each carbon dioxide bubble will end up in the atmosphere, liberated into the air above you. Each trapped bubble, too, will ultimately finagle its way into the vast open air above.

———

The dismal note of death from scurvy opened this chapter, but it ends on a cheerful one: the enjoyment of food and drink.

SEVEN

Drinking the Orange

A Persian Heaven is easily made;
'Tis but black eyes and lemonade.
THOMAS MOORE, *INTERCEPTED LETTERS*, VI

On a train from Chicago to Philadelphia in 1908, Albert D. Lasker (1879–1952), who had left journalism for advertising, happened to meet Cyrus H. K. Curtis, owner of the *Saturday Evening Post* and the *Ladies' Home Journal*. Curtis recommended that Lasker contact Claude C. Hopkins, who had just devised a highly successful campaign for Schlitz beer. Lasker hired Hopkins as his chief copywriter. Hopkins would help make Lasker's advertising agency, Lord & Thomas, the biggest in the world of its time.

One of the early accounts landed by Lord & Thomas was that of the California Fruit Growers Exchange (later known as Sunkist). Its growers were facing a glut of citrus fruit, so they began to cut down trees to reduce overproduction. Destruction of fruit trees is indeed the last resort of citrus producers whenever there is a glut on the market and prices collapse. Lasker was shocked, thinking it wrong to destroy the trees. His agency came up with an idea, coining the slogan "Drink an Orange." With his exceptional sales abilities, he sold the idea to Sunkist, and within a few years, customers used to eating oranges also learned to drink the juice from them. Orange juice quickly became popular, and the orange trees were saved.

Lasker's innovation, the selling of juice rather than fruit, was the forerunner of all the bottled and canned juices now on the shelves in supermarkets. But the introduction of orange juice to a broad market would not have been viable before this point. Lasker's brainchild depended on two prerequisites: pasteurization, which had only recently begun to be used on milk, and efficient distribution via interlinked railroads and truck routes.

For the commercialization of orange juice, it was essential that it be made safe by efficient bacterial disinfection. The importance of this requirement is brought home by the occasional horror story. In June 1999, for instance, there were a few hundred cases of diarrhea in Oregon and Washington due to *Salmonella* infections. These illnesses were caused by unpasteurized commercial orange juice traced to a single processor.

Orange juice came along after a major disaster. In 1918–1919, a worldwide epidemic of influenza killed 20–40 million people, of which 675,000 were Americans. (By comparison, the total death toll from World War I was 8 million people, including combatants and civilians both.) The flu pandemic made Americans keen on taking every possible step to protect their well-being. Health-conscious people tend to equate food habits with protection from illness. In the early 1920s, in the aftermath of the pandemic, American parents were intent on protecting their children from disease, so when doctors recommended drinks such as milk or orange juice for flu prevention, they were quick to provide them to their children.

Giving orange juice to children was also seen as beneficial because of its high calcium content. It worked as an alternative or a supplement to milk. Furthermore, orange juice contained the still mysterious ascorbic acid, later to be identified as vitamin C. Lasker and other advertisers could bank on a vague and somewhat irrational belief that orange juice would magically protect its drinkers from germs in the same way that it protected them from scurvy.

Thus, in the early 1920s, orange juice became an integral part of the American breakfast. Pasteurization made it safe to consume a raw fruit juice, which could be substituted for cooked or stewed fruit, such as prunes or applesauce.

A progenitor of agribusiness was responsible for much of the commercialization of orange juice in those early years. The California Fruit Growers Exchange (CFGE) was then headed by G. Harold Powell, who gave the cooperative the structure of a modern corporation, including an outstanding communications network that enabled growers to get instant and constant feedback from the marketplace. The CFGE also provided advice, based on extensive scientific research, for improving the product, and it introduced new horticultural varieties. The CFGE was an influential organization, and Powell was a genius, comparable to Henry Ford in his managerial skills.

Even after Powell's death in 1922, at the age of fifty, the organization to which he had given such impetus continued its forward momentum. As a consequence, the production of citrus fruit in the state of California overtook that of Florida in volume. For many years, in the 1920s and the 1930s, California was producing about 50 percent more oranges than Florida.

During these golden years, overproduction of oranges in California again became a problem. The growers destroyed surplus fruit in an attempt to stabilize the price, burning tons of oranges with kerosene-fed fires. Such actions were considered shocking during the Depression, when many Americans were starving. A literary masterpiece bears lasting testimony to this scandalous action: John Steinbeck, outraged at the notion of this waste when the oranges could have fed the hungry migrant workers camped nearby, devoted one of the most impassioned passages in his *Grapes of Wrath* to this event. He denounced the burning of oranges as symbolic of all that was wrong with capitalism.

———

After the Depression, Florida overtook California in the production of oranges. But let us jump forward to the aftermath of World War II for the next important episodes in the orange juice story. Wars have a deep influence on culture. In 1945, after three years of experimentation with a high-vacuum process to concentrate orange juice, the National Research Corporation (NRC) in Boston developed a flavorful orange juice powder. The NRC had already used the same process for dehydrating and prolonging the life of penicillin, blood plasma, and streptomycin.

The new development came just as the U.S. Army issued an open order for half a million pounds of orange juice powder for the troops. NRC quickly established the Florida Foods Corporation, with John M. (Jack) Fox at its head. A pilot plant had already been built in Plymouth, in the heart of the Florida citrus belt, to test the new process. Fox set out to win the government contract, for $750,000 worth of powdered orange juice, and succeeded.

In the 1940s, shipping of fresh fruit segments and juice to the Northeast from Florida proved so successful that one grower, Anthony Rossi, discontinued production of his fruit gift boxes. Rossi saw the new technology for flash pasteurization of orange juice, introduced in 1954 as a natural extension of his fruit and juice business. He invested in an evaporator to extract and remove water from the juice. This innovation led to the growth of the Tropicana brand.

Rossi and a few others were ushering in another revolution in citriculture, which tapped a new mass market. They drew on several interlinked social events. The postwar years marked the beginning of the baby boom. The parents of all these new babies had themselves been the first generation to have been reared on orange juice with breakfast. Had they become addicted to the sweet-tasting drink? At any rate, it was natural for them to initiate their children in turn. The postwar years also marked the rise of the suburb. Busy commuter families kept their new refrigerators stocked with frozen dinners and, beginning in the late 1940s, with frozen concentrated orange juice.

Tropicana became a market leader in the frozen juice business. To remain on the leading edge, they turned to technological improvements. In the early 1950s, Tropicana innovated flash pasteurization—raising the temperature of the juice for a very short time. The company's glassmaking plant began operation in 1964. By 1997, the company was making 2 million bottles a day in two of the world's largest flint glass furnaces. But Tropicana did not commit itself exclusively to glass containers. It commissioned the American Can Company to develop a waxed paper carton in half-pint, pint, and quart sizes. In 1969, Tropicana became the first company in the citrus industry to operate its own plastic container manufacturing plant.

Vertical integration of this kind became a model for the whole food industry. Meanwhile, data on the health benefits of orange juice continued to pour in. Children easily absorb iron from either apple juice or orange juice. In a group of healthy infants, carbohydrate absorption from orange juice is greater than from apple juice. Folic acid is found in fresh oranges at a level of about 47 micrograms per fruit—about 12 percent of the daily requirement for this growth factor. Freshly squeezed orange juice contains about 19 milligrams of calcium per 100 mL, sufficient for the daily requirement of a growing child. Calcium plays a key role in human growth, acting as a bone builder.

Vitamin C, also known as ascorbic acid, is an antioxidant. It is not the only one in citrus fruits: polyphenols account for about 85 percent of the antioxidant activity in orange juice. Antioxidants are molecules that retard cell aging and help prevent cancer. Citrus fruits contain high concentrations of several classes of phenolic antioxidants, including compounds known as flavonoids, which have been found to inhibit proliferation of a number of different cancer cell lines. Citrus flavonoids, such as hesperitin from oranges, help to prevent breast cancer.

There is also evidence that orange juice can lower cholesterol levels. Orange juice fortified with plant sterols is effective in reducing low-density lipoprotein (LDL) cholesterol. At the end of 2003, the Coca-Cola Company started marketing such a drink, which it named Minute Maid Premium Heart Wise. A daily drink of orange juice may also reduce the risk of a stroke—an effect I shall return to.

But are all these healthy ingredients of orange juice retained when the juice becomes a commercial drink? Vitamin C is highly oxidizable. Some studies report 70 percent depletion of vitamin C in freshly squeezed juice after seven hours. In the commercial product, a ballpark figure is a decay of 2 percent per day in an opened refrigerated bottle or carton. Industrial processing alters the quality of the juice somewhat. Freezing causes a dramatic decrease in phenols and polyphenols, and thus in the antioxidant properties of the juice.

Like any other food or drink, water included, orange juice can become too much of a good thing. Given its ingredients, moderate use is clearly called for. The acids it contains, predominantly ascorbic and citric acid, can supply too much acidity to the stomach. The sugars contribute to a surfeit of carbohydrates, and their sweetness may become addictive.

Moreover, calcium, if overabundant, puts strain on the kidneys. A glass of orange juice contains about 250 mg of calcium. Fortified orange juice may contain as much as 360 mg of calcium per glass. Our daily requirement is on the order of 1,000 mg. Children need a little more: 1,300 mg. The daily limit for calcium ought to be 2,500 mg, given its connection to kidney damage and kidney stones.

Such concerns apply to children especially. Currently, 11 percent of healthy preschoolers in the United States consume at least 12 fluid ounces a day of fruit juice. Such excessive consumption can lead to obesity because of the high caloric content of the juice. It can also lead to nonorganic failure to thrive and stunted growth. Four ounces of fruit juice a day is the limit for children under two. Otherwise, the juice fills the stomach and the child loses interest in other foods.

There are other, admittedly minor risks associated with citrus juice. One of them is absorption of pesticides with which the fruit has been treated. To mention just one such chemical of widespread concern, about 3 percent of the thiabendazole (TBZ) sprayed on oranges gets into the juice. Various agencies in Europe and in the United States are in the process of setting an upper tolerance limit between 10 and 300 ppb (parts per billion) in orange juice.

—————

Should we then replace the orange juice in our diet with other citrus juices? Grapefruit juice is good for you. This blanket statement has been shown, in the last few years, to be simultaneously true—for both healthy and sick people—and false—for patients suffering from a number of specific illnesses. Like other citrus juices, grapefruit juice provides vitamin C and has a pleasant taste. But any plant or herbal remedy carries a cornucopia of chemicals, some good and some bad. Grapefruit juice does not consist only of water, vitamin C, and sugars. It contains other chemicals as well, such as bergapten, quercetin, naringenin, and bergamotin, to name only a few. Some of these chemicals can interact with drugs administered by physicians.

Bergapten and bergamotin, for instance, deactivate an enzyme in the small intestine, an action that prevents some medications from being destroyed before they get into the bloodstream. This is true of HIV-1 protease inhibitors used to treat AIDS, and it applies to some calcium channel blockers as well. Intake of grapefruit juice enhances the action, as well as the side effects, of such drugs. Serum concentrations of erythromycin (a common antibiotic), cyclosporin-A (another antibiotic), and sildenafil (better known as Viagra) are all increased by the drinking of grapefruit juice.

Certain other drugs are weakened, rather than potentiated, by grapefruit juice. These drugs include the beta blockers celiprolol and talinolol, as well as fexofenadine and nifedipine. In combination with such molecules, the juice activates a protein whose role is to pump molecules out of the gut wall and back into the intestine. Grapefruit juice thus reduces drug absorption as much as sixfold.

Thus, we have come full circle. Orange juice was a commercial success because it was healthy; however, orange juice and other citrus juices, such as grapefruit, can be unhealthy in various ways. Such is the Janus face of nature. Uncritical reverence for nature and the natural can be hazardous, to say the least.

—————

But let us return to orange juice. I have mentioned the possible presence of TBZ in commercially produced juice. Any agribusiness is run these days like an oil refinery, by engineers with specialized training. There is no basic difference in technical sophistication between the manufacture of gasoline for a car engine and the production of pasteurized milk or orange juice. All these liquids, essential to modern commercial culture in industrialized countries, are complex formulations whose precise characteristics are monitored continuously during the process of their manufacture.

Hence, the making of "100 percent Florida orange juice," as this commodity is known, is quite a bit more involved than simply squeezing oranges and selling the juice. Moreover, the industrial plants juicing the oranges work, if not on a year-round basis (genetic engineering has yet to devise oranges that mature during summer, though I am confident it will), at least for three-quarters of the year.

Orange juice production starts in October with the early varieties, typically Hamlins, which bring the production through the holiday season into January. By that time, midseason varieties such as the Pineapple are ready to be turned into juice. By the end of their availability in February, the late-season varieties, such as the Valencias, take over and carry the brunt of the operation until June. And yet the taste of juice is constant year-round! This constancy speaks to the expertise of citrus juice engineers.

At what point does this process transform a product of nature, fruit, into an industrial commodity, packaged juice? Almost as soon as the fruit is picked. The fruit from the trees is collected in tubs holding about 900 pounds each. The contents of the individual tubs are then collected in container-trailers, which hold about 45,000 pounds of oranges. At that stage, the fruit has become raw material, a bulk input for the processing plant.

The capacity of a typical Florida processing plant used to be about 400,000 tons per year. Currently, plants with the capacity for processing more than a million tons of fresh fruit per year are being introduced under the leadership of companies such as Sucocitrico of Brazil, owned by the Cutrale family (more to come on them). In these newer plants, orange juice flows at the rate of a small river: 330 tons of fruit per hour, 750 gallons of orange juice per minute. The analogy with an oil refinery is apt here.

Agronomic engineers, who sample oranges prior to picking, monitor product quality primarily with two numbers. The so-called degree brix measures the percentage of soluble sugars in the juice—the word "brix" comes from the name of the instrument used for that measurement—while the brix-to-acid ratio measures the key parameter to consumers, the balance between sweetness and tartness. Oranges are picked if and only if they measure up, topping a threshold of about 8.5° brix and a brix-to-acid ratio of at least 10 to 1. The USDA Grade A (the best rating) applies only to a juice above 11.5° brix and with a brix-to-acid ratio between 10 and 20 to 1.

As the oranges enter processing, they are washed and sorted by size into batches. The peel is pricked for extraction of essential oils (used as ingredients in perfumery, as we shall see, as well as in the juice) and then set aside, together with the seeds, after the juice has been extracted. Juice and pulp are separated from peel, seeds, and membrane material in less than a second.

The juice is then either pasteurized immediately, and then advertised and sold as "packaged orange juice," or first concentrated to about 65° brix and then sent by pipeline to tanks for cold storage. Prior to shipping to a juice packager, the contents of various tanks are blended, just as for a port or a Scotch whisky. Citrus peel essences and oils, obtained in the course of processing, are now reinjected. They improve the flavor and the fragrance of the blend.

About twenty-five different chemicals make up the flavor of orange juice: while it is a commodity, it is still a carefully prepared formulation. Orange juice technology has acquired a high degree of sophistication. Flavor ingredients are analyzed, identified, and quantified. For instance, the higher fruity note in a Valencia, as compared with a navel orange, is correlated with greater concentrations of some of those twenty-five chemicals.

Orange juice is meant to be cloudy, since the desired flavor and color depend on the insoluble materials suspended in the squeezed juice. Cloud stability is controlled by careful manipulation of the pectin component of the juice. This complex process requires a balance between two enzymes. Pectin methyl esterase promotes cloud formation by increasing complex formation between pectin and calcium. Polygalacturonase inhibits cloud formation by depolymerizing the pectin before it can form a complex with calcium ions.

Frozen concentrated orange juice (FCOJ) was developed in 1945–1946. In making FCOJ, the juice is first heated to deactivate the natural enzymes that are present. Then water is evaporated from the juice. Then— in the final step that was the basis for the commercial success of FCOJ—a small percentage of fresh juice is added to the concentrate to restore the fresh orange juice flavor. The final product is about three times more concentrated than fresh juice.

During the 1979–1980 season, 7.8 million tons (173 million boxes) of Florida oranges were used to make FCOJ, more than 30 pounds of juice per American. But consumers surprised the producers. FCOJ is now in commercial decline, eclipsed by the huge success of chilled, reconstituted ready-to-serve orange juice, especially the calcium-fortified and high pulp variants. More recently, between 1990 and 2000, the consumption of not-from-concentrate (NFC) juice increased threefold. Because of the difference in volume, it is much more expensive to ship NFC juice than FCOJ. In 1999–2000, the relative amounts drunk in the United States were 267.7 million gallons of FCOJ, 661.4 million gallons of reconstituted chilled juice, and 629.9 million gallons of NFC juice. In 1999, Americans drank an average of 5.7 gallons of orange juice.

Processing engineers keep a record—just as they would with an expensive wine or a whisky—of the analytical characteristics of each batch of orange juice. The people monitoring juice quality—U.S. Food and Drug Administration (FDA) inspectors, for instance—use the same kind of information, as well as chromatographic traces and isotopic signatures of the place of harvesting in the juice. They also make sure that the product conforms to its labeling.

An orange appeals to the taste because it combines the sweet and the sour. The balance of the two is what makes citrus fruit in general attractive. A successful grower will provide customers with citrus that is not only juicy, but whose juice also strikes the proper balance. It is a matter of fine-tuning the acidity as compared with the total sugar content. As soon as oranges found a mass market, the need arose for individual growers to gauge this ratio. Could they be provided with some objective measure on which to base decisions about their crop: to harvest, to select, to reject (though discarded fruit was useful nevertheless, as a source of citric acid and pectin), to graft, to hybridize?

In the early 1930s, Glen Joseph, a scientist at the California Fruit Growers Exchange, got in touch with an old friend from college, Arnold O. Beckman. At that time, the CFGE controlled three-quarters of the California citrus production. Its higher-quality fruit was sold under the Sunkist trade name, a label that has endured. Joseph wanted to measure the acidity of the fruit. However, sulfur dioxide was used as a preservative, and its presence precluded the most common method of measuring acidity, with litmus paper, which it bleached and rendered useless. Joseph had also tried measuring the pH—a conventional measure of the acidity of any aqueous medium—with the standard glass electrode. However, the ensuing electric currents were too small for routine measurements. True, this hurdle could be circumvented by recourse to a larger glass electrode, but this made the apparatus too fragile and too bulky for routine field work.

Beckman was then a young assistant professor in the chemistry department at Caltech. After college, Beckman spent two years (1924–1926) at Bell Laboratories, learning electronics, statistical quality control in manufacturing, and how to run corporate research and development. Following his meeting with Glen Joseph, Beckman attempted to design an apparatus that might answer the growers' needs. To measure the intensity of the electric current, he replaced the galvanometer that Joseph had been using with a pair of vacuum tubes that functioned as an amplifier. The result was a device called the pH meter, which was both accurate (± 0.02 pH unit) and rugged. In October 1934, Beckman and his coworkers applied for a patent. In the spring of 1935, Beckman turned the garage enterprise he had started the previous year into National Technical Laboratories (NTL), a manufacturer for the new instrument. Beckman and his wife Mabel displayed the pH meter at the national convention of the American Chemical Society in San Francisco, where it met with great interest but some resistance to the high price tag ($195). Arnold and Mabel launched upon an ambitious national sales tour, in the course of which they won over Ed Patterson, Jr., of the Arthur H. Thomas Company, based in Philadelphia and nationally known. Patterson predicted a 600-unit market over ten years, and he included the new Beckman apparatus in his company's catalog.

Demand turned out to be much higher than forecast, and NTL, which moved to a larger location on Colorado Boulevard—Pasadena's main street—sold 87 instruments in the last trimester of 1935 alone. In 1936, the first full year of production, NTL sold 444 Beckman Model G pH meters. In 1939, Beckman resigned his academic position at Caltech and became a full-fledged instrument manufacturer.

Seen in hindsight, the phenomenal success of Beckman's pH meter might have been anticipated. There are so many practical uses for a pH meter. To mention only a few food- and drink-related applications, the instrument became a crucial piece of equipment for wineries and breweries, bottlers of soft drinks, processors of dairy products, and managers of water treatment plants. And Beckman's little box (12″ wide, 8″ deep, 9″ high), weighing only 15 pounds, was portable—you could carry into the field like a briefcase by its leather handle. It required no prior scientific training and was singularly easy to operate; for instance, the protective lid cleverly switched off the power when it was shut.

———

Beckman's pH meter, daringly innovative in its recourse to electronics, nevertheless was steeped in a tradition more than a century old. The idea of science coming to the aid of the economy, agriculture, and industry had been important since the beginning of the nineteenth century. The Royal Institution had been established in London with that explicit aim. When it recruited the young Humphry Davy as professor of chemistry, that brilliant and dashing scientist did not content himself with outstanding work in pure chemistry—discovering new elements. He went into applications, too. He looked into means to improve agriculture with chemicals, and he invented a safety lamp for miners. Later in the century, the German chemist Justus von Liebig showed a similar attitude in his work. His book on agricultural chemistry was a best seller and was influential for many decades.

This belief, typical of nineteenth-century faith in progress, that science could improve the common lot and help people in every walk of life, was all the more congenial to Americans, who saw it as a reiteration of Jeffersonian ideals as well as the legacy of the lay scientist Benjamin Franklin. The land grant colleges that started sprouting up in the last quarter of the century, with their explicitly agricultural and mechanical mission (still apparent today in names such as Texas A & M) embodied this attitude.

———

A word of caution: "Freshly squeezed orange juice" is only as good as the hygiene of the people making it. As with any food in any restaurant with a careless kitchen staff, too often it is a recipe for contamination by *Salmonella*. Spoiled orange juice can be injurious to the health, but fraudulent juice is another thing entirely.

An analytical chemist from the University of Saskatchewan, Nicholas Low, spent a sabbatical year in the late 1980s at the University of Reading in the United Kingdom. Practically out of idle curiosity, he decided to examine the composition of fruit juices sold in the markets.

The results were interesting and impressive. Seventy percent of British apple juice contained Jerusalem artichoke, used as a sweetener. As for orange juice, a major company from Israel, with worldwide sales, was producing about 25 percent more orange juice in weight than it had oranges. In the United Kingdom, out of twenty-one commercial orange juices sampled in 1990, sixteen were found to be adulterated. Low's results were confirmed by official government statistics.

American companies weren't far behind. A Midwestern manufacturer, it was later shown, had defrauded customers of $45 million over twenty years by adulterating orange juice. Another company netted $2 million in a couple of years.

The Sun Up Foods scam netted that company between $10 and $20 million. The company was not a minor player on the citrus juice fields. Its sales increased spectacularly from virtually nil in 1984 to more than $57 million in 1989. In 1990, when news of the scandal broke, it presented itself as "the largest blender of orange juice in North America and one of the largest in the world."

"Blender" was a well-chosen word. In 1990, an employee went to the FDA and revealed that the company had installed a contraption in a hidden room that pumped liquid beet sugar into the orange juice it was processing. Stainless steel pipes hidden in the walls were set up to appear to be part of the sewage system. In the event of a government inspection, the sugar-carrying line could be turned off, and the outside pipe closed to conceal the illicit sugar pipeline.

A simple calculation makes it easier to understand the temptation to tamper with natural orange juice. The profit motive is obvious. Juice is priced between $5 and $6 per gallon (according to May 2002 figures), and Americans drink an average of six gallons of orange juice per year. This translates into a $5–$10 billion market. Let us assume that a company has cornered some 10 percent of that market, and that its sales are then about $600 million. Its outlay (fixed costs of every description) cannot exceed half that figure, maybe $300 million. If cheating will save only 1 percent of the total cost, it nevertheless amounts to an additional gain of nearly $3 million. With such an incentive, the urge to fleece customers can be overwhelming.

How is orange juice adulterated, and how can adulteration be detected? The main adulterants are corn syrup and beet sugar, since 98 percent of the total soluble organic content of the juice consists of sugars, predominantly sucrose, glucose, and fructose. Corn syrup and beet sugar provide these sugars at a lower cost. However, they cannot simply be poured in, along with the appropriate amount of water to increase the volume of the juice. This would be far too easy to detect.

Instead, the illegal addition of sweeteners is performed with invert sugars; that is, sugars whose sucrose has been cleaved, chemically, into glucose and fructose. The chemical reaction then proceeds until the glucose:fructose:sucrose ratios match exactly those in the juice: approximately 1:1:2. Addition of partly hydrolyzed beet sugar, which is very low priced, has the further advantage of escaping detection by techniques based on ratios of carbon isotopes.

But how do you prove that an orange juice sample is fraudulent? A variety of techniques have been devised to finger the culprits. They all take advantage of chemical signatures. Any natural extract, such as orange juice, shows a characteristic distribution of its chemicals—of *all* its chemicals, including those present in only minute amounts.

Among such telltale ingredients, adulterated orange juice contains D-malic acid. Normally, only L-malic acid is present; the finding of its mirror image, D-malic acid, points unambiguously to fraud. Other revealing chemicals are oligosaccharides, normally present only at very low concentrations, and some amino acids.

Moreover, the precise hydrogen isotope ratio in the sugar molecules amounts to yet another signature. Sugarcane, sugar beets, and oranges produce sucrose by different metabolic pathways. This difference shows up in the isotopic content of the ensuing sucrose molecules.

———

So far, I have described only the production of Florida orange juice. The other major production area, which outdoes Florida in volume, is Brazil, to which I now turn. José Cutrale, Jr. (1926–2004), found his name on the list of the 500 wealthiest people in the world every year, according to *Forbes* magazine. It was relatively new money, and Cutrale was, worldwide, the emperor of orange juice.

The son of a Sicilian immigrant to São Paulo, Brazil, Cutrale was a wholesale dealer of oranges in the central market of that city in 1962. During the night of December 12 in that year, citrus production in Florida was all but destroyed by frost. This was Cutrale's finest hour. He told himself, and others, that Paulista production (i.e., that of the state of São Paulo) could substitute for Florida's—and he was proved right. At about the same time, Cutrale was allying himself with the Coca-Cola Company, one of the reasons for his spectacular success. The Atlanta corporation purchases essentially all of its orange juice from Cutrale's company, Sucocitrico, which it sells under the brand name Minute Maid.

In 1987, new devastating frosts hit Florida, pushing up the price of orange juice to $2,000 per ton. Cutrale consolidated his grip on the frozen concentrated orange juice market, and Brazilian concentrated orange juice became the world leader. Cutrale was astute as well. When the United States and the state of Florida slapped tariffs on imported

Brazilian orange juice and started legal proceedings for dumping against Sucocitrico, Cutrale moved in: he now owns orchards and processing plants in Florida—in Polk County, in particular.

Cutrale's company, Sucocitrico Cutrale S.A., now owns 45,000 hectares of orange groves. Its share of the global frozen concentrated orange juice market is one-fifth. Another awesome statistic: taken together, the states of Florida and São Paulo account for 90 percent of the world production of orange juice. Sucocitrico owns its own tankers for exporting orange juice concentrate; one of these, the Orange Blossom, has a capacity of 13,000 tons.

The orange groves in the state of São Paulo are at the leading edge of today's agribusiness. They employ 90,000 workers: Brazilian oranges are still handpicked, since labor there is very cheap. The orange orchards within the interior of the state are also backward in terms of irrigation. As elsewhere in Brazil, only about 3 percent of the orchards benefit from it. The citrus belt in São Paulo covers 320 municipal districts. The revenue to Brazil, in exports alone, amounts to $1.5 billion (and the Cutrale family is worth $1 billion). While mechanized fruit picking has not made inroads yet, other technologies are state-of-the-art. The Cutrale plantations total 6.7 million trees. Among them are clones, which are reproduced by grafting. As Joaquim Teófilo Sobrinho, who directs the Citriculture Center in Cordeirópolis, which supplies the clones, declares, "All the trees have two mothers but no father." As a result, all the trees are of identical lineage, both resistant to disease and highly productive.

Thus, due to José Cutrale and his company, Brazil is the leading producer of orange juice concentrate (with 60 percent of world production). The family business is now run by his son, José Luis (1946–). Cutrale attributed his success to hard work ("Never take a holiday," he said). He was extremely secretive, and he ran his business with a tight fist. The people in Araraquara, where the company headquarters are located, referred to the barbed-wire-fenced buildings, whose grounds are patrolled by armed guards, as "the concentration camp."

––––––

It is interesting to compare the psychological perception of orange juice in Brazil and in the United States, for the comparison conveys some interesting cultural differences. In the Latin world, that of Brazil, citrus juice became associated with industriousness, cleanliness, and, as unlikely as it may seem in the tropics, with an image of the faraway European Alps and their glaciers.

But we must first look at Swiss immigration to Brazil, which started in the early nineteenth century. In the mid-1810s, the Portuguese government made a contract with the authorities in the Swiss canton of Fribourg. In 1818, the first Swiss contingent of settlers arrived from that canton. This group, made up of 1,700 individuals—entire families in some cases—established a colony, which they named Nova Friburgo (New Fribourg). Not surprisingly, they chose to settle in a mountainous area, in the basaltic coastal range near Rio de Janeiro, about a day's ride on horseback from Rio. The purpose of the Portuguese authorities in organizing such European immigration (formulated in a decree on May 16, 1818) was to populate Brazil with colonists, who would counter the overwhelming African influence that slaves had brought with them from Africa, and would also supply much-needed farmers and artisans, providing an important new impetus for an economy otherwise overwhelmingly based on colonial plantations.

This worked out very nicely. In their enclave, the Swiss colonists reproduced their homeland as much as they could. Their influence was still detectable more than a century later. In the mid-twentieth century, well-to-do inhabitants of Rio would escape the summer heat by driving up into the mountains, two or three hours away, on weekends. In the 1950s, when my family lived in Rio, we would visit Nova Friburgo regularly. There we experienced the eerie feeling of finding ourselves back in Europe. The area around Nova Friburgo featured meadows and cows and Swiss chalets—probably complete with cuckoo clocks. Dairy farming and cheesemaking were economically important activities at the time.

I returned there more recently, just a few years back. With the huge expansion of Rio de Janeiro (a million people in 1950, ten times as many fifty years later), Nova Friburgo has become a satellite city. The small town of the 1950s, with a few thousand people, has now increased to several hundred thousand, and it has become industrialized, producing textiles—lingerie in particular.

———

Anyway, whenever you stop for refreshment in Brazil, you can ask for a glass of *limonada suissa*, the literal translation of "Swiss lemonade."

LIMONADA SUISSA

Two medium-sized lemons
1 liter refrigerated water

A half-dozen ice cubes
12 tablespoons granulated sugar (Brazilians have a sweet tooth!)

* Remove the lemon peel and seeds
* Put the fruit in a food processor or blender, together with the water, the sugar, and the ice.
* Blend well and strain. Serve immediately.

Of course, this is nothing but standard lemonade, if of an ultrasweet kind. What about its name, the adjective "Swiss" specifically? In Brazilian culture, the stereotypical association of Switzerland is with snow-capped mountains and glaciers. It is an ice-cold drink. Swiss lemonade helps counteract the stifling, asphalt-melting summer heat in Rio and other Brazilian cities.

Two uses of lime juice epitomize Brazilians' cultural attitude toward citrus fruit. When I lived in Rio—in the Urca district, just below Sugarloaf, across from the beach at 44 Avenida João Luis Alves—a Brazilian woman, Magnolia Gomes de Oliveira, was the most important person in our household. Formally a maid, this fine person was the self-appointed intermediary between our family of foreigners and Brazilian society.

Magnolia came from a large family of many sisters. Engaged to a firefighter, with whom she went out, mostly on weekends, Magnolia saved her wages in anticipation of being married and having her own home.

She was extremely patient, very even-tempered, and smart, and she became a member of our family. In addition, Magnolia was a born problem solver. Whenever there was a crisis, large or small, Magnolia would remain calm, and she would instantly come up with the best solution.

We also owe to her our introduction to Brazilian foods and cooking. Here is one of our favorites, which she always prepared to matchless perfection.

AVOCADO WITH LIME JUICE

2–3 medium avocados
1/3 cup lime juice
1 cup granulated sugar

* Mash the avocados, which should be ripe, into a pulp with a fork.
* Gradually add the lime juice.
* Let the mixture rest for half an hour in the refrigerator, covered. After this time, the acidity from the lime will have digested the avocado fibers, turning the mixture into a smooth, fluid cream.

- Add sugar to taste. It makes the yummiest of desserts.

Note: The mashing can be performed with a blender or food processor, rather than by hand, as Magnolia did.

Yes, this dish is closely related to Mexican guacamole, but it is sweet instead of being salted or flavored with a chili or Tabasco sauce. Some people improve this dish further with whipped cream, or turn it into an ice cream by adding four whole eggs, 3 cups of whole milk, 1 large can of sweetened condensed milk, 1 cup of whipping cream, whipped, and 1 13-ounce can of evaporated milk. (This mix is first simmered over medium heat, stirring occasionally, for 6 minutes or so, prior to working in the avocado-lime-sugar mixture and refrigerating.) Other people place this mixture in a crust to make an avocado-lime pie.

———

Allow a word from a chemist. Not only is this a delicious cream—to me, it will always be associated with Magnolia's angelic and benevolent smile—it is distinctly healthy as well. Avocados, rich in potassium, carry an abundance of fats, predominantly unsaturated and polyunsaturated. Their proteins sport a majority of acidic side chains—glutamic acid and aspartic acid residues—which helps to explain why they mix so well with sugar and lime juice. Lime juice, by its acidic nature, prevents the browning of avocado pulp when exposed to the air. Besides vitamin C and citric acid, it contains hesperidin as its major flavanone (i.e., as a phenolic antioxidant). It even provides, at parts per million levels, traces of vanillin, the molecule responsible for the aroma of vanilla, which so many of us crave, in ice cream and elsewhere.

———

The sidewalk vendor, a feature of both Anglo-America and Latin America, points to both a colonial past and mostly African roots. The contemporary version is still available in the form of newsstands at street corners and vendors selling lottery tickets or snacks from hot dogs to ice cream, sodas, and fruit juices, a common sight in Latin America.

One sidewalk business common in Brazilian towns is a seed stand. Very often these stands are the dominion of Afro-Brazilian women from Salvador de Bahia (known as the *Bahiana*, pronounced Bah-yanna). This is grassroots capitalism in action. Being a street vendor is also the first step on a social ladder for poor newcomers to the city.

Plate 1. A tiny selection from the cornucopia of citrus fruits: two kinds of ruby grapefruit, navel oranges, blood oranges, Seville sour oranges, sweet limes (the yellow ones), key limes, sour limes, kumquats.

Plate 2. Francisco de Zurbarán (1598–1662), *Still Life with Lemons, Oranges, and a Rose.* Norton Simon Museum, Pasadena, CA. Painted by Zurbarán in the manner of a *bodegon*, its meaning remains elusive. Reproduced by permission from the Norton Simon Foundation.

Plate 3. Joris Hoefnagel (1542–ca. 1600), Flemish manuscript illustrator, *Sour Orange, Terrestrial Mollusk, and Larkspur*. Tempera colors and gold paint on parchment. The J. Paul Getty Museum, Los Angeles, CA. Hoefnagel conflated flowers, fruits, and insects (and other animals too) in a surprising and rather surreal manner. Notice the harmony of the illumination and the calligraphy.

CITRUS DECUMANA.

Plate 4. The shaddock, fruit and tree limb. From Berthe Hoola van Nooten, P. Depannemaeker, and C. Muquardt, *Fleurs, fruits et feuillages choisis de l'île de Java: Peints d'après nature* (Bruxelles: C. Muquardt, 1880). Hoola van Noten had followed her husband to Jakarta. After his death, left with debts to pay and a family to support, she managed to get her splendid neo-baroque book published. Missouri Botanical Garden Library.

Plate 5. A *caipirinha*, synonymous with Brazilian hospitality, is made from sugar, crushed limes, and *cachaça*, a rumlike alcohol distilled from fermented sugarcane. The sweet and sour taste is addictive, but the high alcoholic content renders intake of more than one or two drinks hazardous.

Plate 6. Key lime pie, which originated in Key West, Florida, is made from the key lime, *Citrus aurantifolia* Swingle. The recipe for this American delight dates prior to refrigeration; hence, it uses canned milk, and originally, the pie was not baked. The citric acid would set and thicken the egg yolks, responsible for the pale yellow color. Cooking is recommended, though, to avoid Salmonella poisoning.

Plate 7. The orange tree. Botanical print from F. E. Köhler's *Medizinal-Pflanzen in naturgetreuen Abbildungen mit kurz erläuterndem Texte*, vol. 1 (Gera, Germany: herausgegeben G. Pabst, 1887). This chromolithograph was prepared from the drawings of the artists L. Müeller and C. F. Schmidt. The careful delineation of detail is admirable. Missouri Botanical Garden Library.

Plate 8. Giovanna Garzoni (1600–1670), *Still Life with Bowl of Citrons,* late 1640s. Tempera on vellum. The J. Paul Getty Museum, Los Angeles, CA. The painter was an Italian woman employed by the Medici to paint botanical and zoological specimens. A calligraphy book, illuminated with insects, birds, flowers, and fruits, published in 1625, made her reputation. Henceforth, her work would fetch high prices. She excelled at depicting the folds, the twists, the warps, and the start of decay in natural objects. This piece partakes of the *natura sospesa* (suspended nature) genre, in vogue during the late 1640s.

Plate 9. The "Orbit" crate label, from the 1940s, influenced, one may conjecture, by Clyde Tombaugh's discovery of Pluto in 1930.

Plate 10. Oranges cut in halves to display the contrast between blood oranges and ordinary oranges. The former owe their richer pigmentation to a colder climate. Typically, blood oranges come from the mountains of Sicily. Photograph by elly millican.

Plate 11. A street scene during the pitched mock battle of the oranges in Ivrea, Italy. Like other Carnival events, it upends the usual rules of behavior. Destruction of the once highly valuable citrus fruits partakes of a ritual of derision. Similarly, in the small Belgian city of Binche, the richly costumed Gilles, mounted on stilts, bombard onlookers with oranges from baskets during Carnival. © Alex Masi / Grazia Neri.

Plate 12. Geneviève de Nangis-Regnault (1746–1802), botanical print of the orange tree, *Citrus aurantium* (1774). Nangis-Regnault was a talented painter and engraver whose work combines strength and gracefulness. She provided illustrations for the books written by her husband, Nicolas-François Regnault (1746–1810), an amateur naturalist. Their wife-and-husband team awas not uncommon in the eighteenth century. It evokes that of the Lavoisiers: Madame Lavoisier illustrated the *Traité élémentaire de chimie*, the revolutionary masterpiece written by her husband.

Plate 13. Aerial view of a Florida citrus grove. Worldwide, these plantations share the same regimented look.

Plate 14. The Orangerie at Versailles. This view from the inside of the structure reveals the beauty of its simplicity and functionality. Photo RMN / © Droits réservés.

Plate 15. Raphaelle Peale (1774–1825), *A Dessert (Still Life with Lemons and Oranges)*, 1814. Oil on panel. National Gallery of Art, Washington, DC. Peale was America's first still life painter. Based in Philadelphia, he traveled extensively and was given to intemperance and high living, which killed him at a relatively young age. Courtesy of the National Portrait Gallery.

Plate 16. The Tour d'Argent restaurant in Paris is located on the top floor of this building. It is famous for its *canard à l'orange*, one of the musts for well-to-do tourists in the French capital. Such combinations of tastes originated during the Renaissance in Italy. In reaction to the sweet taste of medieval cuisine and in response to the Arabic influence, aristocratic gourmets developed a preference for more contrasted food pairings, such as sourness of the orange together with the fattiness of duck flesh. Claude Terrail, the owner of the Tour d'Argent, passed away in 2006.

Plate 17. Thinly sliced citrus is essential to bartenders. These delicate slices remind us of the rarity and high value placed on citrus fruits for centuries.

Plate 18. A display of oranges at a market in Rio de Janeiro. Brazil is representative of the world-wide tropics, where oranges are not orange-colored, but greenish. The synonymy between the fruit and the color orange is thus seen to be an artifact of the European experience. © 2007 JupiterImages.

Plate 19. Containers of oranges in front of a juice processing plant. Nowadays, the Garden of Eden is marketed, and its fruits are transformed into a commodity. © Corbis.

Plate 20. Fresh citrus at breakfast, the prerogative of a small elite for centuries, became available to all in the aftermath of World War I, thanks to the imagination of Albert Lasker (1880–1952), one of the creators of modern advertising who coined the slogan "Drink an Orange." © Getty Images.

The next step on that ladder is hole-in-the-wall commerce, also a common sight throughout Latin America. In Brazil, such an operation is routinely centered around an espresso coffee machine. I have no doubt that the *cafezinho* (kah-fay-zee-nio)—as these tiny cups of strong coffee, which Brazilians down daily by the dozen, are known—originated in such places, perhaps in one of the tiny pedestrian streets such as Rua do Ouvidor, in the old city of Rio de Janeiro. These hole-in-the-wall establishments typically also sell cigarettes and matches, cigars and cigarillos, lottery tickets, chewing gum and candy, small pastries such as *cocadas* (compacted cakes of grated coconut and sugar), aspirin and other common drugs, razor blades, and so on. One can lean over a freezer and fish out seemingly ancient ice cream cones and cups. Another common gadget serves to make fruit juice. A wire basket or a bowl holds two or three dozen sickly-looking citrus fruits, still green in patches. But these fruits are the source of the most divine juices, if you can bring yourself to order one. I would not be surprised if a small street bar in Rio or São Paulo, offering a variety of juices from tropical fruits, was where a *caïpirinha* (pronounced kah-hippie-reegnah) was first mixed and named.

But the *caïpirinha* is much more an indoors drink, offered in hotels and private homes, rather than on the street. Brazilians are extremely hospitable. They show a sense of sharing space, food, and drink with an extended network of family and friends. The ambition of a typical middle-class Brazilian couple is to own a large house with a wing for domestic help and at least one guest room. The plantation house remains the model of a home where such hospitality is provided. It can seem overwhelming to guests, who might in turn take advantage and extend their stay for quite a few days and nights.

When one enters such a household in mid-afternoon, in the company of the father after his day in the office, a young woman will appear and, with your consent, will bring you a *caïpirinha*. She will show up for this purpose only at the appropriate time. And the drink will be perfection.

The main ingredients of a *caïpirinha* are white sugar, lime juice, and *cachaça*. *Cachaça* is a very strong spirit, with a history in festivity and folklore.

Months of effort go into making costumes and floats for Carnival in Rio, accompanying intense psychological preparation. As part of it, "the year's song" is selected. During the weeks preceding the opening of Carnival, radio and television stations relentlessly play this song over and over, until it verges on an obsession. Set to the very simple one-two rhythm of a march, the lyrics—often similarly simplistic—address a topical issue. One year, for instance, the song dealt with the powerful national alcoholic drink:

Vocé pensa que cachaça é agua
Cachaça não é agua não
Cachaça vém do alambique
E agua vém do ribeirão.
(You believe *cachaça* to be water / *Cachaça* is not water / *Cachaça* comes from the
still / And water comes from the river.)

————

Cachaça dates back to the colonial period. On plantations, after sugar
had been extracted from sugarcane, the leftovers would serve as feed for
both animals and slaves. The slaves would let this sweet residue, a mix of
cellulose and sugar, ferment for a few days prior to distilling it. *Cachaça* is
thus a raw sort of rum. While its official Portuguese name is the dignified
aguardiente, everyone knows it instead by its popular name, *cachaça*. It is
an essential ingredient in a *caïpirinha*:

CAïPIRINHA

- Into a wooden mortar, pour about one tablespoon of white sugar per serving.
- Cut into quarters one and a half limes per serving.
- With a wooden pestle, crush these into a mixture of pulp and juice while you add
  half a glass per person of *cachaça*.
- Serve the liquid mixture in glasses, adding crushed ice according to taste and
  decorating with a piece of lime.

Warning: This drink is addictive and can sneak up on you! Personally, I
try not to have more than two. A *caïpirinha* shoots a fix of alcohol into
your bloodstream and into your brain at lightning speed. It is a strong,
intoxicating, absolutely delicious alcoholic drink.

The *caïpirinha* is preeminently Brazilian. It is the quintessential beverage of that country, which is expressed in its very name. Brazilians sub-
scribe to the legend that a *caïpirinha* was first mixed in the "interior"—to
a Brazilian, the "interior" is what the frontier was to an American in the
nineteenth century, and the term usually refers to the state of São Paulo.
Indeed, the name *caïpirinha* derives from *caipira*, which refers to a small
town. The *-inho(a)* ending is the affectionate Brazilian manner of referring fondly to a small thing. A small cup of coffee is known as *cafezinho*, a
child or a pet as *amorzinho*. An American equivalent of *caïpirinha* would be
"moonshine," "boondocksade," "hillbilly rum," or something of that ilk.

The supreme irony is that this most Brazilian of drinks is totally of Asiatic origin. The sugarcane comes from Asia (most likely Polynesia), as does the lime.

But birthplace is not everything: in fact, a birthplace can mean next to nothing. Whenever I sip a *caïpirinha*, I relish its Brazilian quality.

––––––

But let us return from the happy hour to breakfast, in order to compare the mental images that citrus juices evoke for a North American and for a Latin American. I shall do so by way of an anecdote.

In the summer of 1962, my wife Martine and I traveled south of Mexico City toward Yucatán. At Coatzacoalcos, where the continent shrinks to the isthmus of Tehuantepec, where the Atlantic and the Pacific are less than a hundred miles apart, we descended from the bus. It was very late, about 1 a.m., but still extremely hot and humid. Another two tourists came off the bus, a couple from New York City on their honeymoon. We all repaired to a local inn and its hammocks.

Early the next morning, about 6 a.m., Martine and I awoke to a rhythmic noise—similar to the sound of logs being split for the fireplace. We got up and joyfully partook of the freshly split coconuts (their splitting was the noise that awakened us) and the other tropical fruits that were offered. What a treat!

After a while, the two New Yorkers joined us. Their despair was comical. What, no coffee! No *New York Times*! No bacon and eggs! And no orange juice. We tried, with scant success, to convince them of the uniqueness of the breakfast that had been provided, and furthermore, which was offered with utter grace and dignity.

We are creatures of habit when it comes to food. I could have told a very similar story about ourselves—the French and their bread, cheeses, and wine. This is food as culture and culture as second nature.

What this comparison shows is the deep influence of mental images. What the image of Switzerland is to Brazilians, the image of California was, from the turn of the twentieth century, to people living in the Northeast and the Midwest. Orange juice is much more than a drink. Albert Lasker's slogan, "Drink an orange," also meant, in a deeper sense, "Absorb a little of mythical California," with its renowned climate and its easygoing lifestyle. Taking a drink is tantamount to an acculturation.

––––––

Being part of a culture brings with it the all-important sense of belonging. Taking this as a privilege, however, is simplistic. How does one avoid becoming culture-obsessed and culture-blinded, to the point of becoming prejudiced against other cultures, to the point of a knee-jerk aversion to coconut milk?

By stepping back from one's own culture, which is our joint obligation to humanity. This is most easily achieved by being open-minded about other cultures, at least to the point of beginning to understand what they are all about. One can become multicultural without losing one's roots in one's own culture.

Food history provides such a perspective. We should enjoy of a glass of orange juice first thing in the morning without becoming hooked on it. The history that I have sketched above tells us how orange juice became part and parcel of the American breakfast. This development was dependent on a series of accidental occurrences; it was historically contingent.

————

I have devoted some space to the mental images associated with orange juice because advertising was an integral part of both its introduction to and its enduring favor with consumers. Advertising is, simply, the marketing of mental images.

The strength of Albert Lasker's idea was the creation of a heretofore nonexistent need. Lasker was riding a social wave: the introduction of orange juice came at a time of growing urbanization, nostalgia for the disappearing simple life—the idealized log cabin or Midwestern farm—and increasing homogenization of the middle class in the United States. Albert Lasker provided John Doe—the average, representative American—with a *natural* drink at about the same time the Coca-Cola Company was trying to corner the same, potentially huge market with its *artificial* drink. Sunkist and Coca-Cola were both banking on the transformation of individuals into consumers.

Of course, Lasker was not the only American to identify the new social landscape and its marketing opportunities at the turn of the twentieth century. He, like fellow pioneers Thomas A. Edison, George Eastman, and, a little later, Henry Ford, perceived this new truth: the mental image outweighed the actual product. Contrary to the dictates of common sense, the prime purpose of business was to sell the former. Which Lasker summed up with the brand name Sunkist, which is in itself a slogan.

Modern advertising was born with the twentieth century. It presented commerce with new ways to expand. In the case of orange juice, it sold a product endowed with an attractive color, aroma, taste, and texture, as well as a manifold of ideas: put simply, health insurance coupled with a fantasy of a tropical paradise.

This part of the tale of citrus originates in the appealing taste of an orange, when the teeth bite into a segment and the mouth fills with juice, slightly acidic, but not too much, and sweet, but not unduly sweet. Science was solicited to render this delight in a juice, and science complied. And science continues to assist citrus growers not only to optimize the production of juice, but also to sell it. For instance, when on October 6, 1999, the *Journal of the American Medical Association* published a paper from a team at Harvard Medical School showing the preventive effect of orange juice (with the intake of other vegetables) against stroke, the advertising arm of Florida Citrus—the equivalent in Florida of the CFGE in California—pounced on the news and launched an integrated media campaign. The story was covered on the *Today Show*, CNN News, Fox News, and so on. The public relations effort generated more than 400 broadcasts with an estimated audience of 125 million. This publicity helped increase U.S. consumption of orange juice by about 10 percent per capita between the mid-1990s and the present.

———

I'll end this chapter with lemon juice, which has so many uses in the kitchen and elsewhere (for instance, as a secret ink). One such use was a product of sheer genius. The dish, named for the painter Carpaccio, and which he allegedly invented, has more than a touch of the genius of the painter. It might be described as an Italian salad, in which the leaves of lettuce are replaced with leaves of meat—that is, with very thinly sliced raw beef.

CARPACCIO

- First, one places on a very large, flat plate a thin film of olive oil—only the very best cold-pressed ultravirgin oil will do.
- Then one covers the whole surface with the meat, sprinkled with grated parmigiano cheese.
- The last touch consists of a few lemon slices. In a tribute to Carpaccio, one deposits them artistically, if not whimsically.

To enjoy carpaccio, put a slice of meat on your fork and then pick up a lemon slice with it, firmly enough so that the lemon's juice becomes mixed with the olive oil. This makes for perfect seasoning—and, personally, I wave away any waiter who shows up with a pepper mill.

EIGHT

Extracting the Essence from the Peel

De la naranja y la mujer, lo que ellas den.
(From the orange and from the woman, take what they have to give.)
SPANISH PROVERB

When he finally made it to the hospital, Bill Winston (not his real name), an otherwise healthy New York City bartender in his mid-thirties, was in agony. After work, his hands were always red and irritated, and then they started itching. When he got home late at night, he would apply a soothing lotion, but his condition worsened. Both hands developed painful sores and blisters. He had tried over-the-counter topical medications of various kinds, but none of them had helped. When Winston went to the emergency room at Columbia Presbyterian Hospital, he was promptly referred to the dermatology unit.

The department head, and his associates, ascertained that the patient was allergic, not to the juice of the fruits he squeezed at work (oranges, lemons, and limes), but to their skin. Winston was given corticoids and told to stop working. Within a couple of weeks, his hands were much improved. After a desensitization procedure, he was able to resume work. However, he henceforth followed the recommendation of his physicians and always wore surgical gloves.

Winston's allergy was induced by chemicals present in the skin of citrus fruit. Repeated contact with citrus peel is indeed a well-known occupational hazard for bartenders. In summertime, when they make cocktails outdoors and squeeze limes in the sun, they often get eczema on the first and second fingers. This contact dermatitis is caused by the combination of sunlight and irritant chemicals. The usual offenders are D-limonene, geraniol, and citral, molecules from the family of terpenes. Indeed, the physicians at Columbia Presbyterian found that Winston had reacted to geraniol and citral, which they proved with skin patch tests.

The peel of citrus fruit contains many chemicals indeed. It provides the cook with a simple way to flavor dishes just as it provides the bartender with flavors for mixed drinks. Outside the kitchen, molecules from citrus peel are prized by designers and manufacturers of perfumes. There is arguably more money to be made from the rind than from the pulp, the seeds, or even the juice. Indeed, discarded orange peels from orange juice plants are carefully saved for later extraction of their essential oils.

But what are chemicals such as limonene, citral, or geraniol? Why does the fruit contain them in its peel? How are they made by the tree? And why are so many of them so useful to us?

————

There is more than one way to skin a cat, as the saying goes. Having never engaged in this particular activity, I have to take it on trust. There is also more than one way to peel an orange, a task that might even serve as a kind of psychological test. Some people, more lackadaisical than others, will absentmindedly use, alternately, a knife or their fingernails, and in removing the peel, will leave no ordered pattern.

Conversely, we all know people who, because they are neat or compulsive, follow a more methodical path. Some first carve the outside in a grid of meridians and parallels before excising small or large geometric segments. There is also the occasional virtuoso peeler, the Houdini of orange peeling, who, holding the blade of the knife obliquely and at a constant angle, starting at one of the poles, will remove a continuous strip of helical shape till the other pole is reached and the orange is left unadorned. Such helical peels are a frequent feature in Dutch still lifes of the seventeenth century, the painter depicting them in hyperrealist, trompe-l'œil manner as an exercise in virtuosity. The viewer is tempted to touch the peel to ascertain whether or not it is real.

Now to a small experiment. Any citrus fruit will do. Take a piece of peel—from lemon, lime, orange, grapefruit, it does not matter. Bend it between thumb and index finger over a piece of paper, and note the dots of oil that spurt onto it; these are the essential oils.

Let us now move from the visible to the invisible. History has revealed considerably more science in the invisible because sensory evidence is both defective and inadequate. By resorting to nonsensory evidence, provided by sophisticated instrumentation together with modern electronics, one elicits the composition of these dots. In other words, one finds out what chemicals are in these oily juices. As a chemist myself, I was trained in this artistry of the invisible.

To sum up the findings, there are treasures within citrus peel. Paradoxically, we ignore them when we discard the most valuable part of the citrus fruit, its skin. The tiny glands it shelters are miniature gold mines. They endow the phrase "golden apples" with rich meaning. The chemicals these oil glands secrete are numerous and diverse, and chemists have turned them into many lucrative products.

For centuries, essential oils from citrus peel were collected and made into highly valuable trade items. They were used as is, without separating out the components. Nevertheless, they were used for perfumes—more on that later—for flavors (orange bitters), and for household formulations, such as waxes for furniture or wooden floors.

————

What exactly does the orange peel contain? The fruit, of the same structure as other citrus fruits—the hesperidium, as botanists call it—has a leathery rind—the so-called exocarp—pitted with pinholes. Each of the holes is the outlet for one of the oil glands. The size of the gland's opening is truly tiny: about 0.3 micrometer. Each oil gland in the exocarp of a citrus fruit rivals a chemical plant: it makes and stores about one hundred different molecules. Furthermore, the production of these chemicals, instead of being constant and invariable, adapts to the plant's circumstances, such as the temperature, the season, the presence or absence of a protective wax on the fruit, and most probably the nature of the insect species on the prowl. Thus, these oil glands can also be compared to a symphony orchestra, as they produce highly elaborate and diverse compositions of essential oils.

————

In the arcane language of chemistry, citrus peel essential oils are known as terpenes. The word "terpene" is related to "turpentine." Indeed, turpentine, which comes from pine trees, is likewise a complex mix of numerous different terpene molecules. Many plants produce terpenes as part of their essential oils (citrus), their resin (conifers), or secretions of their various organs, such as flowers or leaves (citrus again).

Several thousand terpenes are known. Their proliferation is all the more remarkable given that these molecules are all made from a single building block, known as isoprene (natural rubber is a polymer of isoprene). Isoprene has five carbon atoms. Terpenes are known with ten, fifteen, and twenty carbon atoms, corresponding to assembly by and in the plant of two, three, or four isoprene building blocks. The three molecules already referred to as constituents of citrus peel oil, limonene, geraniol, and citral, are ten-carbon terpenes.

Even though scientists have established the structures of basically all the known terpenes, and even though we also know the precise ways in which plants combine them, the reasons for their production by the plant remain obscure. One likely reason is protection against insects. There are probably other reasons as well, but they have yet to be uncovered.

If we make the analogy of isoprene with a syllable in a word, terpenes are polysyllabic words, and there are enough terpenes around for the equivalent of a chemical language. But, unlike Linear B or Egyptian hieroglyphs, this chemical language still awaits deciphering.

———

At home, we can readily flavor a drink or a dish with the terpenes from citrus peel. We can do so directly from the fruit, using various tools devised for this purpose. There are graters with jagged holes smaller than those in the usual multipurpose graters, known as lemon graters. They are perfect for turning a piece of rind into a perfumed dust to enhance a dish. Wash the citrus skin carefully, though, in order to remove the pesticides with which it has probably been treated.

Another tool is the zester. At the end of a stainless steel blade, comparable in thickness to flatware, there is an array of five aligned holes. The zester is ideal for removing thin strips (or zest) from the outer layer of the rind of a citrus fruit. I always use the zester in making the filling for a lemon tart.

There is a third tool for this purpose, the canelle knife (or citrus stripper). Its sharp V-shaped tooth cuts out 1/4-inch-wide strips of citrus peel to both decorate and flavor a drink or a piece of pastry. Bartenders routinely use a canelle knife to zest a lemon or a lime into a mixed drink.

———

We can also incorporate citrus flavors and fragrances into our cooking indirectly through the use of orange bitters. Basically, bitters are mixtures of alcohol and essential oils from citrus. Many formulas exist. Several English firms use Seville orange peel. In New Orleans, there is Peychaud's Bitters, and Abbort's Aged Bitters has been made in Baltimore since 1865.

The best-known bitters, I think, is Angostura, a dash of which enhances a mixed drink such as a Manhattan (one part red vermouth and two parts bourbon, shaken with ice). It owes its name to the town of Angostura, in Venezuela, but it is made in Trinidad. History explains the apparent contradiction. A young, idealistic German doctor, Johann Gottlieb Benjamin Siegert, left his country in 1820 for Venezuela. He was determined to help Simon Bolivar fight the Spaniards for the emancipation of Latin America. Bolivar appointed Siegert surgeon general of the military hospital in the city of Angostura, which in 1846 would become Ciudad Bolivar.

In his spare time, Siegert experimented with alcoholic (45 percent per volume) extracts of blends of various herbs. (The formula remains a secret.) He named the result he came up with in 1824 "Amargo Aromatico." In 1830, he started exporting the product to England. Commercial success ensued. By 1850, Siegert had resigned his Venezuelan army commission to devote himself to manufacturing and commercializing his bitters.

In 1867, his son, Carlos, became a partner. After Siegert's death, Carlos's younger brother Alfredo joined the company. Because of strife in Venezuela, where one dictator followed another, Carlos and Alfredo decided to leave the country. They opted for Trinidad, where another younger brother joined them. To this day, Angostura Bitters is made in Port of Spain, Trinidad.

Angostura Bitters owes its success in part to the name, the felicitous assonance between Amargo and Angostura. Part of the appeal is in the flask: a little brown glass bottle that would be redolent of the pharmacy but for its oversize paper label. Very nineteenth-century in style, inscribed with Siegert's flowery signature and bearing testimony to various awards, it extols Angostura Bitters' health benefits (against flatulence in particular). The label wraps around the bottle all the way up to and past the "shoulders," where it becomes a little crumpled because it's not attached to the glass. This little yellow-capped bottle expresses whimsy, hence its charm.

About the same time Angostura Bitters was being devised in Venezuela, its active principle was being discovered in Europe. In 1828, a man named Lebreton, a pharmacist in Angers, a town on the Loire River in central France, reported his isolation of the bitter principle from oranges, which he named hesperidin. This was a prelapsarian time when the two professions of apothecary and chemist had not yet undergone a split; the fact that "chemist" in Britain means a pharmacy attests to that happier time before the split. Apparently, Lebreton, a contemporary of Siegert, was also engaged in preparing orange bitters. In so doing, he noticed that a powder would occasionally be left behind in the container. When he used a magnifying glass to examine it, he saw "small bodies, in the form of a geode...showing inside a set of crystals based on the circumference of this little geode, and whose ends met at the center."

Lebreton's memoir makes for a fine read. He ascertained that the isolated substance indeed had a bitter taste. He writes repeatedly of the aspect of the crystals, which he terms *une cristallisation mamelonnée*; that is, "a nippled crystallization, each nipple being formed of small needles themselves gathered into arrays isolated from one another, not unlike the arrangement of stamens in Linnaeus's polyadelphia...a radiating but divergent crystallization."

Lebreton also made certain that the substance he had isolated was not an artifact, which might have resulted from alteration of the fruit principles by the alcohol he had used. And he devised several means, using either alcohol or vinegar, to extract hesperidin from the pith of citrus fruit.

––––––

These days, to mention just one among dozens of chemicals, limonene—with its telltale name—extracted from citrus peel is an important raw material. The food and cosmetics industries use it as a source of mint flavor. The pharmaceutical industry uses it as a source of many drugs. But limonene is not a single chemical species. There are two limonenes. They are mirror images of one another, just like our hands. It would be conceptually accurate to refer to them as left-limonene and right-limonene (actually, they are known as (S)-limonene and (R)-limonene). As an analogy, just consider the two words below, which mirror one another:

LIMONENE   ENENOMIL

Words are made of linear strings of letters. Likewise, molecules consist of atoms held together in three-dimensional space. Both types of arrangements, the two- and the three-dimensional, admit of mirror images. Since both limonenes, of each "handedness," are found in citrus rind, it is not surprising that both molecules smell of citrus. However, their odors are different. (R)-limonene smells like oranges, while (S)-limonene smells like lemons.

Handedness is a characteristic property, discovered by Louis Pasteur during the second half of the nineteenth century, in molecules present in living organisms. Limonenes, *both* of which occur in nature, are the exception. In general, only one member of a pair is found in an animal or in a plant. Moreover, since (R)-limonene is an abundant and relatively inexpensive by-product of citrus production—its price hovers around $1.50 per kilogram—it is a choice raw material for the synthesis of organic molecules with interesting properties—drug molecules in particular.

Given that limonene is abundant and cheap, it is a choice chemical for use in the food industry to give a product, especially a beverage, a citrus flavor. This is even more true of citral, the component in lemon oil responsible for its distinctive odor (0.04 ppm is the detection threshold). Accordingly, citral is produced industrially, since citrus peels do not contain large amounts of it. Badische Anilin und Soda Fabrik (BASF) in Ludwigshafen, on the right bank of the Rhine River near Karlsruhe, Germany, is the major supplier worldwide. Its plant has a capacity for production of 10,000 metric tons of citral per year. The problem with imparting a lemon flavor to a drink with citral is that the chemical oxidizes rather quickly in air, which makes the flavor deteriorate. A citral-flavored beverage tastes fresh when it is first made, but off-flavors come in later.

The smell of a grapefruit is unique: there is no way for the nose to confuse it with that of a lemon. The odor of a tangerine differs more subtly from that of an orange. The fragrance of any type of citrus fruit emanates from the whole set of chemicals present in the essential oils given off by the oil glands in the peel. Dozens of whiffs merge into a single fragrance, and yet what we perceive as a smell comes from just a few—often only one or two—of these components. Why is it that, among those chemicals, some show such disproportionate influence? Even though they are minor components in a peel's oil, and are sometimes present in such minute amounts as to be measured in parts per million (ppm), they provide the aroma with its signature.

Mandarins, for example, owe their odor to thymol, N-methyl anthranilic acid (as the methyl ester), gamma-terpinene, and alpha-pinene.

That of Kabosu sour oranges comes nearly exclusively from citronellal. A solution of this chemical in water at the 160 ppm level has the Kabosu aroma. Another Asian citrus, the Hyunganatsu fruit, has an aroma determined by octanol and linalool; extremely dilute aqueous solutions of these chemicals (from the family of alcohols)—a mere 2 ppm is sufficient—give off the fresh and fruity Hyunganatsu scent. The aroma note of grapefruit (which depends on 1-p-menthene-8-thiol and nootkatone) arises from only about 5 percent of the essential oil. The lesson for food scientists is the absolute necessity of determining all the volatiles in an aroma. Going by the numbers alone is not recommended. An apparently negligible ingredient may be the key to a pleasurable, soothing, perhaps even an addictive smell.

As for the generic odor of citrus fruits, it is tempting to blame it on limonene. Not only is this terpene present in the skin of all citrus fruits, but the human nose is highly sensitive to it. We detect it routinely at the 10 ppm threshold. Some sources claim a detection threshold of only 0.21 ppm for (R)-limonene, but linalool, with a value of only 0.0038 ppm, has an even more powerful scent.

————

The economic value of citrus peel and pith comes about predominantly from the presence of fragrant substances within them, which accounts for the gold in the golden fruit.

In 310 BC, the Greek philosopher Theophrastus (ca. 372–287 BC) wrote of laying citrus fruit among clothes to give them a pleasant odor, as we do with lavender, and to deter moths. Subsequently, during the late Middle Ages and the Renaissance, people of both sexes wore pomanders to ward off stenches and foul air. A pomander consisted initially of an apple or an orange stuck with a few cloves or other fragrant spices. Later it became a piece of jewelry, a small hollow container worn around the neck, or from a belt, with similar ingredients in it. The word "pomander" derives from the French *pomme d'ambre*, "apple of amber," in which "amber" refers to ambergris, the perfume ingredient, not to the fossil rosin. By Elizabethan and Jacobean times, the pomander had become traditional wear at Christmas and New Year. As Ben Jonson wrote in a Christmas masque, "He has an Orange and rosemary, but not a clove to stick in it." The Virgin Queen herself always wore a pomander.

These pomanders were wrought from gold, silver, enamels studded with gems, ivory, or porcelain (the latter in the seventeenth and eighteenth centuries). They were shaped like oranges and were hollowed out in the manner of the original fruit.

―――――

Peel from at least a dozen different kinds of citrus fruits continues to be a major source of raw material for perfumery. Bitter orange oil is expressed from near-ripe bitter oranges. The essential oil from bergamots is also cold-pressed from their peel. Likewise for the essential oils from lemons, mandarins, sweet oranges, pink grapefruit, and so on.

To return to the bitter orange, this species is a treasure trove for the perfumer. Neroli oil is distilled in spring from the freshly picked flowers, while petitgrain oil is distilled from young shoots or twigs and fresh leaves. The latter name, the French word for "small grain," relates to this oil's original source, which was immature, hard, green fruit, resembling "grain."

Neroli oil is very expensive; petitgrain oil is easier to come by. The much cheaper petitgrain oil is among the most common adulterants of neroli oil: it gives it a bitter-woody nuance, which the trained nose of a perfumer easily discerns. Neroli oil may also be diluted or replaced with so-called Portugal neroli, distilled from the more plentiful flowers of sweet orange trees, which makes for a thinner, less richly aromatic oil.

These three essential oils (bitter orange, neroli, and petitgrain) are extensively used by the perfume industry. In 1990, when perfume components were officially classified into seven families, citrus oils were categorized as *hespéridés*. Their presence in a perfume, or better yet in a cologne or in an *eau de toilette*, imparts a fresh, gay, and tonic note. Indeed, most *eaux de toilette* carry them in some degree or another.

The same word, "expression," names the process of extracting essential oils from citrus peel and verbal utterances. Why? Because pressing the discarded skins after the inside of the fruit has been eaten or juiced brings out the valuable essential oils. Bruising the skin of a citrus fruit is enough to make the oil ooze.

"Expression" comes from the Latin verb *expremere*, with the meaning "to squeeze out." It thus describes the squeezing of the many essential oils from citrus peel. The task of a writer, not unrelated, is awareness of the many meanings in a word. He or she has to squeeze out not all, but a few, of these meanings—as if a given word were juicy from the numerous oil glands within. Linguists term such multiplicity of significance *polysemy*.

If the "bartender's syndrome," as dermatologists call it, is a contact dermatitis primarily affecting the hands, "writer's syndrome" could be said to affect the tongue and the pen (or the word processor): the sufferer of such a syndrome might have an inordinate taste for the dictionary, for rich, complicated words, and for the multiple meanings a single word may carry. Could it not be said, however, that good writing, in like manner to inhaling a fragrance from citrus, extracts from simple words a whiff of an aroma with which to flavor sentences?

PART THREE

Symbolic Extractions

NINE

Symbolic Meanings of Citrus

Si Dios te ha dado limones, dedicate a hacer limonadas.
(If God has given you lemons, apply yourself to making lemonade.)
SPANISH PROVERB

In Greek mythology, Atalanta had many suitors, but none of them interested her. Under pressure to marry, she relented, but with one condition: She would marry the man who could outrace her. If she won, he would be killed.

An athlete rose to the challenge. Seeking help from the goddess of love, Aphrodite, he received three golden apples from the garden of the Hesperides. He leapt onto the track. As Atalanta drew near, he dropped the first apple. She bent down, picked it up, and he continued to lead. Nimble-footed as she was, Atalanta still gained on him. He dropped the second fruit. She grabbed it and lost ground again. Once more, he increased his lead. When for the third time she threatened to pass him, he dropped the third apple, and he went on to win both the race and Atalanta.

The Atalanta myth evokes not only Aphrodite, in her capacity as goddess of fertility, but also the cult of the chaste Artemis. Three golden apples—arguably oranges—as a kind of offering in order to win one woman: this story is reminiscent of the judgment of Paris, who had to choose between *three* women to present *one* apple to the one he chose. The Atalanta tale is echoed in stories the Greeks told about the Amazons, strong women with no use for men. The race, which embodies a transgression of separation of the sexes

129

in athletics, is a symbolic restoration of the natural order, threatened by Atalanta's insistence on celibacy. This chapter will deal with oranges and with associated deities, in various pantheons, as symbols of two polar opposites: female virginity and female fertility.

By equating golden apples with oranges, we may surmise the Atalanta myth to have been contemporary with the arrival of the first citrus fruit from Asia—a consequence, perhaps, of Alexander the Great's conquests. In a related myth, part of the Herakles cycle, the golden apples from the garden of the Hesperides gave immortality to whomever possessed them. They were thus doubly precious. A connection begs to be made with alchemy, which around the same time (fourth century BC) was reaching Greece and the Hellenistic world from China and India.

A major aspect of Chinese alchemy was potable gold. Chinese proto-chemists had devised procedures for turning the precious metal into aqueous suspensions of tiny particles, colloidal gold that one could drink. The notion was that the inalterability of the noble metal would be transmitted to whomever drank it, conferring on the drinker good health and immortality.

My conjecture about the near contemporaneity of the Atalanta myth and the onset of Hellenistic alchemy is borne out by the integration of the former into the latter, which can be seen in the classic alchemical manuscripts. The title of one of the most famous, which the German Michel Maier wrote during the early modern period, is the 1617 *Atalanta Fugiens* (or *Atalanta in Flight*).

During Greek antiquity, religion and culture greatly overlapped. Religion extracts symbolic meanings from natural objects, such as citrus fruits, endowing a culture with them. Citrons were important to Jewish religion; oranges were likewise important to Greek mythology. We are the heirs to symbols that are informed by natural history. In this particular case, nature invites the symbolism. Orange trees carry, at the same time, branches of white flowers and of golden fruit. The white flowers connote virginity; the oranges, fecundity.

I turn now to Sandro Botticelli (1446–1510) for an illustration of the enduring symbolic meanings of the orange tree's flowers and fruit. Two of his most famous paintings, *The Birth of Venus* (ca. 1485–1486) and *Primavera* (ca. 1478), prominently display both. In the upper right of the former, the Florentine artist has depicted the garden of the Hesperides. The flowering orange orchard is gold-limned: tree trunks are highlighted with diagonal gold strokes; dark green leaves are edged in gold and each given a gilded spine; petal tips are dipped in gold. The goddess of love, rising from the waves, is about to enter the orange grove, where the flowering trees symbolize the birth of love.

Venus is present in the earlier painting as well. She stands amid an orange grove. *Primavera* was painted for the marriage of Lorenzo di Pierfrancesco de Medici to Semiramide Appiani. Orange trees are shown—Botticelli was a keen observer of nature—in both blossom (chastity) and fruit (fertility), as befitted the union of a virtuous lady with her noble husband.

Throughout Christian art and literature, white orange blossoms are an attribute of the Virgin Mary. Such symbolism carried into the Victorian age. In nineteenth- and early twentieth-century France, a bride would carry a bouquet of orange blossoms, later preserved under a glass globe in the nuptial bedroom. The symbol of her prior virginity, it stood there as a harbinger of her wished-for fertility.

———

Such symbolism explains a little-known episode in Iberian military history, the so-called War of the Oranges. This conflict between Spain and Portugal deserves renewed attention. Without it, the present world remains opaque. Why is it that the foreign policy of Portugal continues to be dictated by knee-jerk hostility to whatever political proposal Spain supports? Why is it that Portugal pushes for annexation of the Spanish province of Galicia? Why is it that Spain has been so successful in advertising its claim over what it considers territory illegally or unfairly occupied by Britain—namely, Gibraltar—whereas Portugal's similar claim, against Spain this time, to the city of Olivença—Olivenza to the Spaniards—is conveniently ignored?

This story has a small cast of characters: a ruthless emperor, a weak king and his vain queen, a detested prime minister—and a bunch of orange blossoms.

The time was August 1800. The ruthless emperor was Napoleon. France was at war with England. Spain was allied with France, and Portugal with England. Portugal had been effectively under the domination of the British since the end of the seventeenth century. If Napoleon contrived to detach Portugal from its British alliance and occupy it militarily, it would be a significant defeat for the archenemy of France. Hence, the emperor pushed for Spain to take over Portugal and throw its support behind him. He sent one of his most trusted generals, Berthier, as his special envoy to Madrid. His express mission was to "excite by all possible means Spain to go to war against Portugal." At the same time, Napoleon expressed his determination to annex Portugal as part of France.

Now to the weak king: Charles IV, king of Spain, was a Bourbon, of the same family as Louis XVI, the king of France, who had been guillotined in 1793. After the execution, Charles IV went to war against revolutionary France, but more as a symbolic gesture of solidarity than as an all-out military effort. By the time Bonaparte, the future Napoleon, came to power, the two countries had made peace and had become allied against Britain. Charles IV, fascinated by Napoleon, was unable to resist his will.

Furthermore, Charles was under the thumb of the queen, Maria Luisa, and of her former lover, Godoy, the royal favorite. Manuel Godoy y Alvarez de Faria (1767–1851), a member of the Spanish petty nobility, owed his brilliant political career—he was already a minister at twenty-two, and he would stay in power for nineteen years—to his passionate affair with Maria Luisa. Even after the affair ended, he retained control over the queen. As the French ambassador, Alquier, wrote home in November 1800 to his Minister of Foreign Affairs: "Godoy allows himself acts of violence and brutality which a soldier would not dare on a prostitute."

Finally, in the spring of 1801, after dragging their feet as long as possible, Charles IV and Godoy went to war against Portugal. The war was short, lasting only about three weeks, until the city of Olivença fell on May 20. The Spanish invaders and their French allies met with very little resistance from Portugal. The offensive was more of a military stroll in enemy territory than a bona fide war. Olivença was annexed by Spain, and remains Spanish to this day.

Godoy, the commander of the Spanish forces, made a delicate, gracious, and chivalrous gesture, one totally out of character for him. Riding through the citrus groves near Elvas, on the way to Lisbon, he had a bunch of orange blossoms picked, which he dispatched home to the queen. This episode explains why historians call it the "War of the Oranges."

The annexation of Olivenza, whose return Portugal has been calling for over the centuries, explains that country's total lack of sympathy toward Spain's demands for Gibraltar's return, as do the centuries of British-Portuguese friendship. This history further explains Portuguese encouragement of the autonomy of the Spanish province of Galicia—the region of Santiago de Compostela—which is culturally and linguistically much closer to Portugal than to Spain.

———

War has its derisive antidote at Carnival time. Battles of flowers are an integral part of the parades city dwellers enjoy watching, whether in Rio, Genoa, Nice, San Antonio, or San Diego. Carnival celebrations have become events for tourists, and indeed, they can be picturesque. The small town of Binche, in Belgium, draws some of the biggest crowds. Its main festive attraction is the so-called Gilles: giants, people walking on stilts, wearing Inca costumes and headdresses of ostrich feathers. The Gilles carry baskets of oranges, which they toss to onlookers. The Inca costumes may have originated in the sixteenth century, but the oranges are a relatively late addition to this particular ritual. Until the turn of the twentieth century, the Gilles threw nuts and pieces of bread, or sometimes apples. Oranges were just too expensive to be thus disposed of.

The pelted oranges in the Binche Carnival are an import from Italy, where they have been part of the Carnival tradition for centuries. A traveler wrote about Carnival in Rome in 1847:

The third class of missiles, or the Indifferent, vary in the manner in which they are applied. Should they be gently tossed, with a sweet smile, we may safely class them among the Complimentary; but when thrown with violence, they are most decidedly offensive. They consist, for the greater part, of oranges, lemons, large balls of sugar, heavy bon bons, and bouquets in which the stem is the principal part.

Such customs endure. For instance, the Carnival in Ivrea (a small city in the Piedmont, north of Turin, best known as the location of Olivetti's headquarters) features a three-day pitched battle of the oranges, during which the combatants throw the fruit at one another. Among the combatants are riders, some in carts drawn by single or double pairs of richly caparisoned horses; their opponents are foot soldiers. The two camps, or armies, total more than 3,000. The Ivrea battle of the oranges is rooted in history. It reenacts, with symbolic violence, bloody uprisings by the townspeople against their lordly oppressors, Raineri di Biandrate (twelfth century) and Guglielmo, Marquis of Monferrato (thirteenth century).

During this mock battle, violence is fierce. Buildings are protected with netting to shield windows from hurled oranges. Innocent onlookers are provided with highly visible red caps for identification and (relative) safety. And the ripe orange projectiles fly about for hours on end, splattering on bodies, heads, faces, and Carnival costumes; those that have missed their target crash to the ground. By the time the epic tournament comes to an end, the streets and squares of Ivrea show an awesome welter of debris, covered as they are with an eight-inch layer of citrus gore and orange pulp. This rubble makes quite a sight.

One can readily understand the selection of the orange, a solar fruit symbolic of both golden light and opulence, as an emblem of Italian Carnival. The occasion is a festival celebrating the return of light at the end of winter, the burning away of the cold months; it is also a festival of fertility, helping to ensure a plentiful harvest and the fecundity of women.

Carnival occurs at a special time of year. But oranges had a festive role in European towns at other times of the year as well.

––––––

Throughout modern times in western Europe, at least up to the end of the nineteenth century, oranges were symbolic of affluence and good times. As a luxury fruit, in contrast to more ordinary and cheaper apples or plums, they accompanied festive occasions. Served at weddings and at banquets, oranges were also savored at the theater. During Carnival, the special time of year for social mock-upheaval, oranges were thrown about in carefree gestures of aristocratic disdain.

Some individuals were designated orange-men and orange-women, who sold the fruit. These specialized peddlers roamed the streets of large cities carrying large baskets filled with oranges. Orange sellers, a familiar sight in large European cities for hundreds of years (until the end of the nineteenth century, when large central markets were built, rendering them unnecessary), probably originated in the Arab world. As late as the nineteenth century, orange vendors in Palermo, Sicily, advertised their wares with the strange cry, "Here's the honey!" which was very similar to the cry of orange vendors in Cairo, who yelled "Honey! Oh, oranges! Honey!" There is no better explanation for the use of the word "honey" than to advertise the sweetness of a normally sour fruit.

One may surmise that orange peddlers first roamed the streets of most cities around the Mediterranean. Then, about the time of the Renaissance, when travel became safer and trade developed between northern and southern Europe, these orange-men and orange-women appeared in London, Paris, Amsterdam, Hamburg, and other big cities of the North.

––––––

From the Renaissance until the first World War, theaters were a place where one would always find orange sellers. They were the predecessors of the vendors now selling popcorn and candy at the entrance of an American movie theater or hot dogs and chips at a stadium.

During Elizabethan times, when William Shakespeare, Ben Jonson, and Christopher Marlowe saw their plays staged at the Globe Theatre, their audience was munching gingerbread, nuts, and oranges—and audience members were not shy about throwing these items at the actors whenever they deemed their performance unsatisfactory. During the seventeenth and eighteenth centuries, gingerbread and oranges continued to be the traditional fare for patrons of the London theaters. Citrus vendors could be found in the theaters and in the streets of such cities as York and Manchester.

In Elizabethan England, and long afterward, the theater was synonymous with immorality. The reader need not be reminded that both the Carnival and drama originated, in Greek antiquity, with the cult of Dionysus. Festivals of the god were drunken revels. They were times—and I am thinking also here of the Roman Lupercal—for sexual indulgence, for a frenzy of debauchery, when men attacked and raped women, among the lower social classes especially.

This kind of behavior was perpetuated by theatergoers in the Elizabethan age and afterward. Gentlemen attending a London theater would get drunk on wine, beer, or gin. They would become rowdy and randy, making for the nearest female. Often, these were the flower-girls or orange-women—who may have welcomed the attention as an opportunity to rise above their lowly condition if a real connection were made. To go from peddling oranges to becoming a courtesan or a mistress to one of the wealthy gentlemen was a desirable and seemingly attainable goal.

The following incident, which occurred on December 20, 1700, documents the lifting of inhibitions on the part of theatergoers and the habitual abuse to which orange-women were subjected:

Sir Andrew Slanning, having made a temporary acquaintance with an orange-woman, while in the pit at Drury Lane playhouse, retired with her as soon as the play was ended, and was followed by Mr Cowland and some other gentlemen. They had gone but a few yards before Mr Cowland put his arm round the woman's neck; on which Sir Andrew desired he would desist, as she was his wife. Cowland, knowing that Sir Andrew was married to a woman of honour, gave him the lie, and swords were drawn on both sides.

One may also recall that King Charles II had Nell Gwynn as a mistress, a former orange-woman herself. Until the end of the nineteenth century and the time of Sarah Bernhardt, the theater and the opera continued to be scenes of amorous encounters and liaisons between wealthy men and working-class women, who were thus socially elevated.

So, if grapes are the fruit of Dionysus, they are not the only one. Oranges, if we are to trust the evidence of Carnival celebrations and boisterous theatergoers, were another fruit bearing the imprint of the Greek god. And that of several other Greek divinities, Aphrodite and Artemis among them.

———

As his eleventh labor, Herakles (Hercules) sailed to the garden of the Hesperides and brought back three golden apples. His twelfth labor, not unrelated, was to bring a cornucopia, which the Hesperides had filled with the golden fruit, to the underworld. Heracles presented this gift to Hades, who had abducted Persephone, Demeter's daughter. In much the same way, antiquity had a celebration equivalent to our modern American Thanksgiving. Greek matrons celebrated the Thesmophoria, a ritual of fertility and bountiful harvest addressed to Demeter and Persephone.

The European custom of oranges and tangerines as gifts at Christmas and New Year derives from such a tradition. The holiday season, which falls in the dark of winter, is a solar ritual of renewal and hoped-for fertility of the land and people alike. A related European holiday custom is wearing a pomander, as mentioned earlier.

The cornucopia had a durable cultural impact, which I can vouch for. My father and I arrived in Rio de Janeiro from France in July 1950. The whole city was unnaturally quiet. Rio was still in mourning after the shock of the national team's loss to Paraguay, just a few days earlier, in the final game of the World Cup of soccer—Coupe Jules Rimet, as it was then called.

I was a young teenager, and this was my first visit to the South American continent. The first sights and smells and sounds of this exotic city made quite an impression. We stayed in a hotel, the Excelsior, conveniently located in downtown Rio, just a couple of blocks from the building in which my father had his office. At that time, this hotel combined luxury and Old World charm with comfort. Only a few years later, such palatial abodes would be made both obsolete and redundant by the American-style hotels that started rising, as they did in Honolulu and Miami, along the length of the Copacabana beach.

When we checked in—it may have been mid-afternoon—very few details registered. I was in a daze, to some extent because of the time difference, but mostly from physical exhaustion. We had been on an Air France flight from Paris, with refueling stops in Madrid, Dakar, and Recife, for a solid 26 hours.

My first distinct recollection is of awakening the following morning to a waiter serving breakfast in the room. He was carrying on an immense platter the Garden of Eden.

Of course, there was coffee, together with freshly baked croissants, buns, pastries, butter, and jam. But to me, the most astounding feature was the cornucopia of tropical fruits the waiter brought in. There were two or three types of freshly squeezed juices. There were slices of watermelon. There was a whole pineapple—my very first acquaintance with that fruit. There were, as usual in Brazil, at least three or four different varieties of bananas. There was papaya. There was mango. There were several types of oranges. And maracuja. And goavas. And goiabada.

From this experience, I know exactly how the European navigators of the eighteenth century must have felt when being showered by the natives of a Pacific island with welcoming gifts of the most appetizing local and natural products. They must have felt they had just arrived in Paradise or the Enchanted Isles. To this day, the myth has endured. It stems from the navigators' descriptions of arriving in the New Land—in Polynesia, for instance. They sent back alluring narratives of their voyages, with descriptions of the local flora and fauna by their resident naturalist, illustrated by their resident artist, also a part of the discovery team.

––––––

The cornucopia on our breakfast table at the Excelsior reminds me of the Dutch still lifes of the seventeenth and eighteenth centuries. These paintings convey the same sense of wonder at the sheer luxuriance of nature, at the diversity of shapes and colors and tastes of tropical fruits. Such works brought home to the Netherlands' cold and gray climate the marvels that Dutch sailors were encountering. The practice of horticulture, as an attempt at re-creation of these tropical and equatorial scenes, also flourished in the Netherlands at that time. Flowers came to the fore, since under the Dutch temperate climate and even in greenhouses, they prospered more easily than fruit. And of course, a whole genre of bouquet paintings also blossomed from the Dutch school in that period.

While the Dutch were roaming the oceans on their impressive intercontinental forays, other nations were not idle. The Portuguese played an essential role in the European expansion. The Spaniards were shipping home gold and silver from the Americas on their galleons. Meanwhile, the Portuguese launched a triangular trade, which the British, the French, and the Dutch quickly joined. They took on board slaves from the West African coast, shipping them into forced labor on the plantations of the West Indies or Brazil, and hauled the products of those plantations—such as rum and sugar, bananas and citrus—back to European ports (such as Bristol, Lisbon, Nantes, or Bordeaux).

Slave labor left a durable imprint on the Brazilian population, especially given the Portuguese practice of miscegenation. Its African component gave Brazilian culture many practices—evident in its religious and culinary traditions—with strong influences on daily life. Citrus trees, a tropical bounty, became linked symbolically to the luxurious sensuality sometimes attributed to Afro-Brazilians.

———

How is this history linked to Brazilian food history and to current food traditions? The answer goes back to the colonization of Brazil by the Portuguese. Much of Brazilian culture is a legacy of life on the plantations that thrived in the Pernambuco area in the northeastern part of the country during the colonial period, from about the eighteenth century onward.

The Portuguese started settling Brazil in the sixteenth century. Single men were sent to that country, awarded by their king huge captaincies and absolute power over the land and people they ruled. These often petty tyrants subjugated the indigenous Indian tribes. They took full advantage of the women offered by the Indians as a hospitality ritual. They exploited the men as well, as forced labor. They initially grew mostly sugarcane and citrus. Coffee was introduced in the 1770s, and cocoa about a century later, when the Portuguese government began to understand the true economic potential of the Brazilian colony. Its value came not so much from Brazil's natural riches (gold, silver, diamonds, vegetable dyes, and so on) as from its agricultural production.

Portuguese immigrants carried with them their notion of society, learned from the Portuguese nation itself. They implanted in South America a similar mosaic or patchwork society. Portuguese culture was a blend of three main currents: besides the old Roman and Christian element, there was an agricultural and fishing component of Arabic origin, heavy on citriculture—the Moors had introduced groves of lemon, orange, and tangerine trees in Portugal—and with expertise at processing sugarcane. Another legacy of the Arabs was their naval prowess. Their lessons were well learned: within a century or two after the demise of al-Andalus, Spaniards and Portuguese started their voyages of discovery to the Indies. The third main influence derived from the Jewish contribution to the Portuguese population and culture up to the sixteenth century.

Portuguese Jews were well-educated professional people, lawyers and physicians, bankers and accountants, professors and high-level administrators. Thanks to the Inquisition, this last segment of the population contributed a disproportionate share to the colonization of Brazil: numerous Jews and New Christians (Jews who had converted) escaped to the New World.

Brazilian society consisted of the Portuguese colonists, the local Indians, and the imported African slaves—500,000 by 1700, 2.5 million a century later. From these elements, the Portuguese built a plantation system. The Portuguese lord—the *coronel*, as he came to be known in Brazil—built a huge stone house—a palatial abode, not a mere reproduction of the family home in the mother country. Next to this big house, known in Brazilian as *a casa-grande*, were shacks for the slaves. The stereotypical image of the lord of the house holds that he spent his time idle, resting in his hammock, eating and drinking, dallying with his concubines, and shouting orders to the numerous servants.

The Portuguese in Brazil were weaned from their traditional diet. Arrival in Brazil had both enriched and impoverished their food supply. There was no more milk, cheese, or meat. The land they owned was devoted not to raising cattle—an iffy proposition anyway, given the climate and the endemic parasites—but to a monoculture, sugarcane. Hence, not only was the Portuguese colonist deprived of protein from animal sources, but he was forced to substitute new foods: manioc instead of wheat; corn instead of the vegetables grown in Portugal. The colonist was also introduced, of course, to the cornucopia of tropical fruits, most of which had been totally unknown in the mother country. Like British colonists in North America, who after a century accustomed themselves to local foodstuffs, especially to eating corn, using maple syrup instead of honey or sugar, eating a lot of succotash, and so on, Portuguese settlers gradually adopted manioc (the *farofa* flour), corn, black beans, bananas, and pineapples. Their main non-indigenous foodstuff was citrus, to which we shall return.

———

Brazil was first settled on the Atlantic coast. To this day, most of the population occupies a relatively narrow strip along the ocean. Hence, seafood offers a protein complement to the diet. However, Portuguese and native Brazilians differ in their attitudes toward fishing and in the ways in which they prepare and serve fish. The Portuguese see fishing as Mediterranean people do. The Brazilians have a much more African attitude. The attendant culture in its Afro-Brazilian expression brings us back to the duality of Aphrodite/Artemis.

Cuisine appeals to the gourmet and has a marked erotic component: food is not unlike sexual gratification in this way. How, then, do Brazilians like their food to be cooked? The cooking of fish is an excellent example.

Loup au fenouil, to me, exemplifies Mediterranean cuisine. The chef brings the fresh catch to the table for inspection and admiring approval. The fish is then removed, prepared, and grilled over embers. It is scented with a few sprigs of fennel inserted in its incised belly. Other branches of fennel are thrown on the fire as the fish, caressed with olive oil, is grilled on both sides.

This dish epitomizes the excellence of Mediterranean cuisine, which makes the elaborate and the sophisticated appear as if it had been improvised, as if the ingredients had been thrown together on the spur of the moment, maybe with divine inspiration. In any truly Mediterranean culture, from Collioure to Palermo, Mykonos, Malta, Cyprus, Tunis, or Algiers, this dish is emblematic of the legacy from our hunter-gatherer forefathers.

Loup au fenouil makes use of fennel, a wild plant native to all the shores of the Mediterranean, together with its giant look-alike, the ferule (a fertility symbol in the Dionysus cult). In the mindset of hunter-gatherers, no one can predict what each day will bring. *Loup au fenouil* allies the fisherman's catch with the anise flavor of the herb, collected by women. Grilled *loup au fenouil* has the taste of great open spaces, swept by salty winds.

Contrast this typical Mediterranean dish with almost any fish preparation devised by Brazilian natives. One is mentally prepared for Brazilian cuisine, genetically related to the Portuguese, to be steeped in Mediterranean tradition and style. But it is subtly but significantly different. The African component, with flavors that at first seem strange to a European or a North American palate, dominates. The Brazilian attitude toward the sea is totally different from the Mediterranean one, and that difference permeates the cuisine.

All around the Mediterranean, the sea is like a member of the family. Like a mistress, it is unpredictable. But it has been charted, sailed on, fought with, and harvested for millennia, not centuries. The *loup* that a Provençal fisherman hauls from the deep is associated with familiar figures from mythology, goddesses such as Amphitrite or Aphrodite, basically benevolent and friendly to mortals.

Fishermen are a vital segment of the Portuguese population. Daring and sturdy, they crossed the whole North Atlantic, fetching cod from Newfoundland. They have bravely persevered for eons with sailboats akin to the caravels of Columbus and Vasco da Gama. In the process, they have established small colonies down the North American coast, along the shores of Massachusetts and Connecticut in particular. To this day, cod remains the basic Portuguese foodstuff—probably one of the main supplies of protein.

To Brazilians, the Atlantic Ocean is definitely not a member of the family. It is a dangerous no-man's-land. Watching Brazilians playing happily on the beaches, riding the breakers or playing volleyball, it's easy to get the wrong idea. This playfulness is very much restricted to the beaches. The open ocean, the unknown turquoise expanse beyond the breakers, is a threat. Swimmers fear it, thinking it infested with sharks. Brazil is a country of planters and plantations, not one of fishermen who thrive on marine riches. Brazil has very few fishermen for such a major country. There are, correspondingly, very few fishing ports. The Brazilians, generally speaking, fish from the beach, whether they haul in fish in nets thrown into a lagoon next to the beach or embark in boats from the beach.

In spite of Portuguese being the language of Brazil, the country has not become a Portugal of the tropics. Fishermen, in fact, are a powerful rebuttal of this simplistic analysis. Brazilian fishermen do not operate like European fishermen who undertake long voyages, but more in the tradition of Native Americans. Especially in northeastern Brazil, in states such as Salvador and Ceara, they go out on rafts known as *jangadas*— their sailors are known as *jangadeiros*. These are their fishing vessels. They embark on these exotic-looking contraptions, very much unprotected from the waves that sweep across their decks tens of miles from shore, and are often gone for several days. Their courage matches that of the Portuguese fishermen.

SEA BASS WITH TANGERINE JUICE

1 lb sea bass filet (or any comparable fish with a delicate white flesh)

1 teaspoon crushed garlic

2 tablespoons salt

1 cup tangerine juice

1 tablespoon olive oil

1 teaspoon star aniseed

1 pack of aniseed (or coriander)

2 tablespoons rice vinegar

2 teaspoons soy sauce

1/2 onion

- Oil an ovenproof earthenware or cast iron pan just large enough for the fish, and set the oven to 400°F.
- Sprinkle salt on the fish, season it with the spices (except for the pack of aniseed), and pour the oil over it.
- Decorate the fish with thin slices of onion, carrot, and tangerine.
- Pour the tangerine juice over the preparation.
- Bake uncovered for 20 minutes. Remove the fish carefully from the baking pan.
- To the cooking juices, add the vinegar, soy sauce, and the pack of aniseed and bring it to a boil.
- Pour this sauce over the fish and serve. Makes four servings.

This recipe features tangerine juice, and not as an afterthought. It is central to imparting to the bass flesh an opulent taste, both tangy and sweet. It gives the beast the flavor of the cultivated land. It melds the fish with the nectar of fruit grown on plantations. In so doing, Brazilians—who, as we have seen, have always amounted to a population of settlers mostly on or near the coast—remove a natural product into the realm of arboriculture. They thus provide, as in so many other aspects in Brazilian life, a syncretic creation.

———

Ethnic diversity may lead to the coexistence of various cultural traditions, or it may result in a melting pot. The latter applies to Brazil. The world was made aware of its ability to integrate various traditions by the 1959 movie *Orfeu Negro* (Black Orpheus), which transcribed the tale of Orpheus and Eurydice to Rio's shantytowns. In the 1950s, the French theater director Jean-Louis Barrault was struck, during a tour in Brazil, by the resemblance of some of the Brazilian rituals of *candomblé* and *macumba*, imported from Africa, with some of the features of theatrical plays in ancient Greece. He thus incorporated these elements into his staging of Aeschylus's *Oresteia*, for which Paul Claudel provided the translation.

Such a convergence of Afro-Brazilian rituals and Greek mythology hints at the importance of myth in Brazilian life and customs. Immortality, conferred by the golden apples of the sun, is one of these potent beliefs.

———

It is always worth scratching the surface of the everyday and the apparently mundane. By digging deeper, one may learn unexpected facts.

Take, for example, the unsurprising finding that a number of small Brazilian cities, especially in citrus-producing areas, hold a yearly Festival of the Orange, complete with the election of an Orange Queen and two princesses.

Two such cities are Taquari and Montenegro, in the southern state of Rio Grande do Sul—RS for short, the official postal abbreviation. While most of the production of citrus in Brazil is concentrated in the interior of the most industrialized state, São Paulo, some also occurs in RS. It is located in two valleys, those of Rio Taquari and Rio Cai. The city of Taquari is on the shore of the eponymous river, while that of Montenegro is in the Cai valley.

Colonists in RS, many of them immigrants from Germany, worked for the most part—except in the larger cities such as Pôrto Alegre, where they controlled international trade and business—as farmers in the countryside, establishing small farms with an average area of about 25 hectares. They became active citrus growers. During the period 1825–1850, the Taquari orange was established in RS, and it was shipped out through the aptly named Pôrto Alegre.

Even the election of the Orange Queen assumes deep significance. It goes back to Old Norse mythology, of the same type that Richard Wagner dipped into for his Ring of the Nibelungen. One of the Germanic deities, the goddess Idun, was married to Bragi, the god of poetry. Idun was the custodian of the golden apples of youth. Just like mere mortals, the gods in the Germanic pantheon were threatened with aging. However, by partaking of the golden apples, they could retain their youth forever. The very name Idun means "The Rejuvenating One."

It should thus come as no surprise that both Montenegro and Taquari choose an Orange Queen every year. She stands for Idun, representing the force of life, the elemental sensuality and the oneness with nature that are key elements in Germanic ideology. The cultural inheritance of the settlers persists; they have found a genuinely Teutono-Brazilian way of expressing that inheritance.

TEN

Images of Citrus in Prose

Vistoso como naranja para muestra
(Conspicuous like an orange for display)
A PHRASE COMMONLY USED IN ARGENTINA

At given times of the year, some people go through a ritual that involves dressing in orange-colored garments.

First, fans rooting for the Dutch soccer team, whenever it is engaged in international competition, wear orange T-shirts. Their cars and their buses display panels, signs, and banners in the same color. The orange color in itself indicates allegiance to the national team.

Likewise, in Northern Ireland, the aptly named Orangemen, Unionist militants of the Protestant faith, who stand against the Catholic Irish and the unification of Ulster with the Republic of Ireland, stage defiant and colorful marches through the streets of Belfast and Londonderry in commemoration of the Battle of the Boyne, which William III of Orange, King of England, won in 1690 against James II.

In the United States, Princeton University alumni, during reunion weekend at the end of the school year, don gaudy costumes of orange and black, the school colors. They also stage a parade, known as the P-rade.

These social rites share a joint historical origin: a pun. The ruling family in the Netherlands are the Orange and Nassau. The first part of this dual name stems from their possession, during medieval times, of the small principality of Orange, in southern France, not far from Avignon. The city there is known, to this day, as Orange, after a Celtic toponym. But when oranges came into the area, it was fortuitous that the name of the fruit and the name of the place were the same. The Orange and Nassau took the color orange as their official color. As mentioned earlier, the name of the fruit has an Arabic, not a Celtic, origin. Nevertheless, the shared name made for a durable association between the color orange and the Dutch royal family.

––––––––

This chapter will document other wordplay involving citrus. Generally, to name is to select, hence also to omit. Consider bird names: the name "cardinal" refers to the color of a bird (the male, actually), not to its size or eating habits. A roadrunner is named for its motion on the ground. A woodpecker owes its denomination to its foraging. Thus, to name a thing is akin to extracting one trait from a group of many.

But there is a catch: the same thing may have different names. Any language, such as English, harbors sublanguages such as slang, baby talk, the technical languages of professionals, craftsmen, and scientists, and so on. Live circuitry interconnects these variations into a single language, however. Take citrus seeds—"seeds" itself is a word at the borderline between ordinary English and the language of science.

Parents brainwash their children into belief in two kinds of language, sacred and profane, or clean words and dirty words. This is nonsense. There is just one language. To illustrate this, peel a tangerine and remove the seeds from the segments. What is their other name? Pips, correct? And where does the word "pip" come from? Is it to be adjudged a clean word or a dirty one?

Any dictionary gives the answer. Pip is short for "pippin." The latter is one of numerous English words of French origin (from 1066 and all that). A citrus seed, in French, is *pépin*. Pépin, in turn, comes from Latin. In Latin, *pipinna* is the name for a little boy's prick: the duplication of the p sound emphasizes the tininess of his appendage. Other words, in French, English, and many other languages, describing the bodily functions of this organ are related. For instance, "penis" and "pip" are cognates. An orange pip was thus named because, improbably, it reminds one of a child's penis.

Coming back to the French *pépin*, a French expression for getting into trouble is *avoir un pépin* (literally, "to have a pip"; with a distant echo of the swearing of a Roman, referring to the minuscule size of the private parts of his antagonist in a quarrel). One might speculate that when English speakers exclaim, "this is the pits," they ought to have said "this is the pips"—the mutation of the final consonantal sound, from p-initiated into t-initiated, is easily explained: "pip" and "pit" are what linguists term a minimal pair, with the smallest possible number of distinctive features in their enunciation.

————

Names are also coined in the other direction, from citrus to body parts, as when a woman's breasts are compared to grapefruits, lemons, or oranges. A suggestive painting, *Naranjas y limones* (*Oranges and lemons*), by the Andalusian painter Julio Romero de Torres, shows a woman naked holding a few oranges close to her chest.

Let us move on to a group of metaphors that come from the squeezing of the juice from a citrus fruit.

After the end of World War I on November 11, 1918, the Allies exacted payment of reparations from Germany. The Treaty of Versailles made provisions for the payment of a bundle of money by the loser to the winners—to the French and the English in particular. We know, in historical hindsight, the consequences of humiliating Germany in this way: economic upheaval, which caused the Weimar Republic to collapse and allowed the rise to power of the National Socialists. Rearmament, Hitler, and the dreadful rise and fall of the Third Reich ensued.

Far from imagining such consequences, England and France wanted to bring proud Germany to its knees, and to make it poor. They wanted to bleed it lifeless, to turn it into a squeezed, discarded lemon—which, incidentally, is where I believe the expression "a lemon," denoting a nonfunctioning car, originated.

The evidence is a couple of sentences from a jingoistic speech that a British politician, Sir Eric Geddes (1875–1937), gave at Cambridge, England, on December 10, 1918, almost exactly a month after the end of the war:

The Germans, if this Government is returned, are going to pay every penny; they are going to be squeezed, as a lemon is squeezed—until the pips squeak. My only doubt is not whether we can squeeze hard enough, but whether there is enough juice.

————

Let us consider the naming of the colors of citrus fruits, which owe their colors to bright natural dyes and pigments. I have often visited New England when, under cover of darkness, Jack Frost does his yearly paint job. A recent autumn trip took me to southern New Hampshire, to the area of Durham and Portsmouth. The trees gave the illusion of glowing, so rubicund had they suddenly become. Oaks, maples, birches, ashes, willows, sycamores—with evergreens providing a standard green reference—all competed to be worthy of a photo spread in the pages of *National Geographic.*

The autumn leaves owe their vivid colors to the same carotenoid and anthocyanin pigments that citrus fruits, as well as many other vegetables, such as carrots and tomatoes, contain. Some trees are still lime green. Others have put on their seasonal lemon-colored suits. Some are tangerine-hued. Others are a bright orange. Still others give themselves the russet coloring of a kumquat.

Language is a living entity. It enriches itself constantly with new words. In so doing, it retains a core. Though it is not solid and invariant, its rate of change is measured in centuries, rather than in years or decades.

The names of colors are a case in point. By and large, they were already set by the Elizabethan age. Most of the names of colors in Shakespeare are still familiar to us; that is, they have more or less remained the same. While chemistry and the dye industry started to produce new colors by the thousands in the 1850s, few of the names of these artificial products made it into the common language. They found their specialized niches in technical languages, such as those of painters, printers, food colorists, and cosmeticians.

The names of colors continue to embody today the encyclopedic view of nature and of the world that prevailed before the New Science of Galileo, Descartes, and Newton. That view was descriptive rather than analytical, relying heavily on analogy and rhetoric, the part standing for the whole (synecdoche). Natural history, at least until early modern times, was the repertory of the wonders of nature: gems, colorful flowers, edible fruits, exotic woods, wild animals, and so on. Witness the enduring names of common colors: the words turquoise, ultramarine, emerald, and amethyst all evoke precious or semiprecious stones. Common flowers gave us lily white, cherry blossom, lilac, violet, and rose. Carmine, cochineal, oxblood, and purple are all names with animal origins. Ebony and mahogany are derived from wood. Fruits of various kinds have been a rich source of names for colors: lime green, apple green, and cherry red associate names of plants with those of a color manifold. And indeed, some fruits lend their name directly to that of a color. Examples include apricot, chestnut, peach, citrine, and, last but not least, orange.

The color orange deserves to be singled out for its ethnocentricity and for its implicit downgrading of countries of the Southern Hemisphere, which deserves some explaining. American English and British English differ slightly in vocabulary and in spelling. The same is true of Portuguese. That language is spoken in the mother country by about 10 million people, and by about 170 million people in Brazil.

Slight variations differentiate the two types of spoken Portuguese. The Portugal Portuguese will tell you gleefully that their version of the language is the more modern. Brazilians have retained quite a few archaic terms from the colonial period as well as a quaint pronunciation—just like the version of the French language spoken in Quebec.

As an example, while the color orange has the name *laranja* (orange) in Portugal, Brazilians term it *cor-de-laranja* (or *cor de laranja*—the correct spelling is a bit controversial); that is, "orange-color."

This is all well and good—except for the incontrovertible fact of the true color of the fruit in the two countries. In Portugal, a European country, with sharply differentiated winter and summer seasons, oranges are . . . well, orange in color. Their peel contains carotenoids, the same kinds of molecules that give autumn leaves and carrots their color. Thus, in Portugal, the word *laranja* denotes the color of the *laranja* fruit people can see on the trees.

In Brazil, a large Southern Hemisphere country traversed by both the equator and the Tropic of Capricorn, even ripe oranges remain green. Because of the absence of seasons in most of the country, especially in the state of São Paulo, where the bulk of Brazilian citrus is produced, the fruit lacks the weather cues that would make its carotenoids overrule the green chlorophyll—as happens in the fall foliage of New England. In Brazil, overripe oranges tend to show a few yellow spots as the balance between the green chlorophylls and the yellow and red carotenoids starts to shift. Thus, in Brazil, the word *cor-de-laranja* denotes a color, which parents teaching the language to their children might liken to that of the carrot—*cenoura* in Portuguese.

———

Now, from the names of colors to toponymy, the names of places. Often, the name of a thing denotes its place of origin. This is true of many items of trade, from wines (Bordeaux, Burgundy, or at a more localized level, Meursault, Gevrey-Chambertin, etc.) to liqueurs and alcoholic extracts (Curaçao and Angostura are two citrus-related examples) and many other grocery items, such as Dijon mustard. Oranges were known initially (according to Theophrastus) as Persian apples or Median apples, Media being located in antiquity in the northwestern corner of modern Iran.

Once again, the converse is also true. Quite a few places (there are many Orange Counties in the United States, for instance) were named for the orange groves they hosted.

Laranjeiras is the name of a district in Rio de Janeiro. Its name is less familiar than those of the beaches—Copacabana, Ipanema, Leblon, for instance—yet it is a rather central part of the city. Driving from the area of the Municipal Theater and Avenida Rio Branco toward Copacabana, after leaving the Morro da Gloria hill to the right, one comes to Rua das Laranjeiras; literally. "the street of the orange groves."

When I lived in Rio—when I was a *carioca* (as inhabitants of Rio de Janeiro are called)—Laranjeiras was a popular shopping area, much lower in social status and prestige than the area of the office buildings along Avenida Rio Branco, the residential areas along the beaches, or those near the base of Pão de Açucar (Sugarloaf), such as Urca, the Yacht Club, and Praça Vermelha. There were then two types of public transportation in Rio, streetcars and minibuses. The latter, known as *lotações*, were public taxis, seating eight to ten people and stopping on call. Many would be seen racing along the route from Ipanema and Leblon through Copacabana to the city center. As for the streetcars, known as *bondes*, they looked like the cable cars of San Francisco. Women and young children sat inside on wooden benches. Men and teenagers stood outside, on wooden steps running the whole length of the car, where they would hold on to handlebars (or to one another)—which also made for a sportive and picturesque ride. Several streetcar routes followed Rua das Laranjeiras. I took one daily to school to the *Lycée français* of Rio, named *Liceu franco-brasileiro*.

The Laranjeiras district of Rio was named for Rua das Laranjeiras, which led at one time to orange groves on the outskirts. I imagine they occupied an area in between the lagoon and the slopes of Mount Corcovado, upon which stands the giant and downright ugly concrete statue of Christ the Redeemer, with extended arms, now dominating the whole modern city of Rio.

ELEVEN

Images of Citrus in Poetry

La monja
cantaba dentro de la toronja.
(The nun / sang from inside the grapefruit.)
FEDERICO GARCIA LORCA

To name is to select a single aspect of the thing to be named, a restriction that does not, however, translate into words having single meanings, except in technical jargon and scientific language. The word-thing correspondence is not univocal: it is at least equivocal. It may even open up, accordion-like. Poetry builds on the resonance of words, on their multiple meanings.

Wordplay is often involved. Light poetry makes use of humorous wordplay. Serious poetry does not forswear wordplay—far from it, it only makes the poem's meaning deeper.

Connected to humorous verse, wordplay is a major resource for writers of advertising copy. Journalists use it too, as an attention grabber. For instance, the French language sports magazine *Planète Foot* specializes in soccer. When the Dutch national team, in its orange jerseys, was competing in the 1998 World Cup, the magazine called an article "*Des oranges à déguster sans modération*" (An Orgy of Oranges), referring to the enjoyment the public had watching the orange-clad footballers play. This title played on the phrase "*consommer avec modération*" (consume in moderation).

Poets often assemble words that sound alike. Consider, for instance, Goethe's rhyme-pairing of the two German words *blühn* and *glühn*. The linguists term these a minimal pair, like their equivalents, "bloom" and "gloom" in English. In his famous poem *"Kennst du das Land?"* (Do you know the country?) from *Wilhelm Meister*, Goethe writes,

Do you know the country where the lemon trees bloom
Where the golden oranges glow among dark leaves...

In this poem, Goethe (1749–1832) tells of his love for Italy and its conspicuous lemon and orange groves, which he had recently observed on a 1786 trip. The German writer followed in the footsteps of earlier travelers to the Mediterranean from northern Europe, Dutch and Flemish painters such as Pieter Brueghel (ca. 1525–1569)—think of the Italian seascape in *The Fall of Icarus*. Writers traveled there as well, such as Charles des Brosses (1709–1777), president of the Dijon Parliament.

Goethe's poem tapped a rich vein. In France, Gérard de Nerval (1808–1855), translator of Goethe's *Faust*, became similarly enamored of the scents and sights of the Bay of Naples (as revealed in the *Chimères* sonnets). Likewise, Alfred de Musset (1810–1857) wrote of the Bay of Genoa and the lemon groves there in *"La nuit de septembre."*

———

English poets also drew on the exotic connotations of citrus trees and fruit; witness Byron. His poem "The Island" speaks in perfect pitch about voyages and their woes and about mutiny and other ocean ordeals, but also of the South Seas and the exotic fruit found there. The poet describes their unrivalled, luscious taste:

Held the moist shaddock to his parchéd mouth,
Which felt Exhaustion's deep and bitter drouth.

The shaddock, a fruit from faraway lands, could not help but appeal to an enthusiast of the British Empire like Rudyard Kipling. "Frail" is the name for a large basket holding figs, raisins, and other fruits grown in sunny, exotic climates. Thus runs a lovely line of Kipling's "Rhyme of the Three Captains" (1890):

He has stripped any rails of the shaddock-frails and the green unripened pine

with its rush of assonances (rails/frails, and unripened/pine).

———

The Spanish poet Antonio Machado (1875–1939) recalls his childhood perceptions in the poems in his collection *Galerias*, in which he refers to his memories as soap bubbles borne away in the wind. One such memory is of *"il limonar florido,"* the flowering lemon tree.

The contemporary South African poet Don Maclennan has found a voice sober and yet sensuous, as in the powerful simile

Winter sunlight, clean as a cut orange.
A stone wall breaks the wind.

Federico Garcia Lorca (1899–1936) went one step further in his surrealist poem *"Vals en las ramas"* (Waltz in the branches). He hung a poem on its rhymes. Sound now overrules meaning:

La monja
cantaba dentro de la toronja.
(The nun / sang from inside the grapefruit.)

Garcia Lorca also likened the plight of the writer who has lost his inspiration to a dry, fruitless orange tree in *"Cancion del naranjo seco"* (Song of the barren orange-tree):

Librame del suplicio
De verme sin toronjas.
(Free me from the ordeal / of seeing myself fruitless.)

———

Closer to Goethe's wonder at the golden fruits of the garden of the Hesperides is the Greek poet Odysseas Elytis (1911–1996), as in "Drinking Corinthian Sun":

I thrust my hand into the wind's foliage
The lemon trees sprinkle the summer pollen.

Poetry can be aphoristic and incantatory when it partakes of the sacred. Turning to prose and to the profane, as contrasted with more ritualistic expression, there can be poetry in folk wisdom too. *Finnegans Wake* includes in an episode of feasting, general rejoicing, and dancing "with a great deal of merriment, hoots, screams," a proverb of Joyce's invention, "so sure as there's a patch on a pomelo." The phrase Joyce came up with is structured like a popular saying, complete with alliteration.

Wallace Stevens (1879–1955) remains the poet of citrus par excellence. One of his best-known poems, "Sunday Morning," gives ecstatic expression to longing and pining for the simple life, a sentiment not uncommon in the modern age. We aspire to a sense of belonging, to feeling at ease with the planet—in short, to a reunion with the gods of paganism.

The poem opens with an image of domestic ease and bliss:

Complacencies of the peignoir, and late
Coffee and oranges in a sunny chair

But this first stanza—the 120-line poem, written in 1914–1915, in unrhyming pentameter, has eight stanzas—rushes into an evocation of death. With a jolt, Stevens recalls the image of the oranges, affixing it, surprisingly, to a dirge, to a funeral procession:

The pungent oranges and bright, green wings
Seem things in some procession of the dead.

To me, this is reminiscent of a Venice burial, the gondolas gliding silently along the Grand Canal as in a scene in Roger Vadim's 1957 movie *Sait-on Jamais* (No Sun in Venice), accompanied by the Modern Jazz Quartet playing their "Cortege" piece.

The poet thus sets up an equivalence between earthly delights and the religious promises of heaven. His poem then continues under a shadow as Stevens asserts that the key to earthly fulfillment is the bite of death, "the mother of beauty," as he writes.

After this invocation of death, Stevens writes a paean to life in the seventh stanza, his poem unfolding to the dance of life—the unmistakable poetic equivalent of Stravinsky's contemporary (1913) *Rite of Spring*:

Supple and turbulent, a ring of men
Shall chant in orgy on a summer morn

Another poem, "It Must Change," presents a full palette of colors, such as the "garbled green" of wild orange trees. Stevens evokes citrus and other fruits—bananas, melons, pineapples—on a tropical island.

In 1949, the Connecticut Academy Arts and Sciences was about to hold its thousandth session. The Academy decided to make it a memorable occasion. It commissioned an original musical composition by Paul Hindemith and a poem by Wallace Stevens—and two scientific presentations as well. "An Ordinary Evening in New Haven," which Stevens wrote for this occasion, is one of his most important poems. He wrote it at the age of seventy, near the end of his life. However understated, the piece is a lyrical testament of sorts. Indeed, the poet later expanded on it for inclusion in *The Auroras of Autumn*, another collection.

"An Ordinary Evening in New Haven"—the title is ironic—consists of near-autonomous individual sections, themselves enclosing assertions in the manner of the maxims by French writers of the seventeenth and eighteenth century, such as Pascal and La Rochefoucauld, both of whom Stevens admired. These aphorisms give the poem an oracular tone, another distinctive trait of Stevens, consonant with his melding of concept and expression.

The first section alludes to the problem of empiricism: should we trust our sensory perceptions? Related philosophical questions crop up: are representations of reality confined to the mind (idealism)? Do words refer to things in the outside world or to other words only? What is the proper place of metaphysics?

In the second section, Stevens gets rid of God, no longer the Alpha and the Omega. To the poet, the real is primordial. Indeed (in the third section), we wallow in reality. Its dominion encompasses everything, religious paraphernalia included, rituals and festivals, calendars of saints, the nonobvious and the invisible too.

In the fourth section, the poet further delineates the duality of the physical, the realm of facts, and the metaphysical, the province of mere language, whose phrases, like used coins, gradually lose their edge and the richness of their meaning.

In the fifth section, the poet as seer enters; he is himself inhabited by the poem. And the poem itself is but an offshoot of reality. The poet, in contrast to his earlier characterization as "untouched / By trope or deviation," nevertheless puts to use such tropes with muffled repetitions, for emphasis and for modulation, as in

resembling the presence of thought,
Resembling the presence of thoughts.

I won't proceed with a detailed analysis. "An Ordinary Evening in New Haven" is a great poem. Ambitious in scope, rich in meaning, it is breathtaking in the majesty of its inspiration, depth of perspective, and seemingly colloquial inner harmonies of language, artfully wrought in great detail, as many commentators have pointed out.

The last section Stevens read on that 1949 evening in New Haven, "In the land of the lemon trees," reminds me of those enchanting prints in old travel books recounting eighteenth-century voyages of discovery, illustrated with woodcuts subsequently touched up and colored by hand, depicting landscapes, flora, and the traveler's encounters with natives.

Besides evoking voyages of discovery, Wallace Stevens plays with words most engagingly in these assonance-laden lines. As he shows, the *elm*/*lemon* permutation interconnects two kinds of trees.

Not only two kinds of trees. There are two lands, two languages, two cultures; this poem moves back and forth between geographies, and between geographies of the mind. The last section of the poem is dualistic in both meaning and sound, semes and phonemes, landscapes and seascapes, the "land of the elm trees" and the "land of the lemon trees." It rings with assonances, as in "dangling and gangling." It resonates with similar sounds, used to highlight contrasts, such as that between light and shade, "clods/weeds" or "blond/bronzed." The latter is a near-repetition with the addition, the second time around, of a voiced sibilant. Further on, Stevens reiterates, to haunting effect, the same phonemic sequences in the same order, /l-n-d/ and /r-n-d, in *land* (twice) and in tu*rned* a*round*.

This playfulness is akin to that in a gestalt switch (shifting from one pattern, or paradigm, to another, totally different one). It is underscored with a sense of cadence and harmony, with melodic patterns, with the intricate balance of the vowels and the patter of the consonants.

In this, the final exhalation, systole- and diastole-like, the lines expand from the formulaic into the pacific. And the poem paints a picture, one of fulfillment. Somehow it connects with other landscapes, also verbal, also pristine and citrine, like those Barry Lopez, for instance, is wont to depict.

But let's turn from the pristine and citrine to the opulence of Dutch seventeenth-century still life paintings.

TWELVE

Fruit as Image

rare fruit of all kinds...to draw from life.
JOACHIM VON SANDRART

At the turn of the sixteenth century, Flanders was among the most prosperous regions of Europe. Its textiles wove wealth into the economy. Flemish dealers imported wool from England. Once across the Channel, it was processed in Flanders. The finished products were sold throughout Europe.

A highly profitable line was the making of tapestries for aristocrats and rich merchants, dignitaries of the Church and monks all over western Europe. Examples of these tapestries, huge decorative pieces covering entire walls, can be seen at the Cloisters in New York (an annex to the Metropolitan Museum of Art) and at Musée de Cluny in Paris. The best-known motif of these splendid sets of tapestries is the unicorn, but a singular motif in these tapestries and many others is the orange tree. There is an orange tree in all kinds of tapestries, whether sacred or profane, whether documenting everyday life or allegorical; whether they were meant for a household, a palace, or a church.

That the orange tree appeared in such artworks and was so much in fashion is evidence of its novelty. Whereas the Italians and the Spaniards had already tended orange orchards for centuries, the tree could not stand the colder climate of the Low Countries. Flemish and Dutch travelers became acquainted with citrus on trips to the South, to the lands of the citrus groves.

Only wealthy burghers could afford to make such trips. And what about the state of the roads? It had taken the economic revolution of the twelfth century to reconstruct or develop north-south European trade routes. The thirteenth and fourteenth centuries were times of consolidation, with a slow, long-term increase in population countered by repeated epidemics such as the Black Death. Toward the end of the fifteenth century, with kings like France's François I leading the way, the palaces of the aristocracy to the north of the Alps and the Pyrénées started to include orange trees.

During the last quarter of the fifteenth century, however, orange trees had yet to reach lands north of the Loire Valley, including Paris and northern France, Britain, and the Low Countries, which at the time consisted of both Flanders and the Netherlands. Even boxed orange trees, sheltered from the night cold, were unknown there.

There is one revealing clue. Master drawings (*cartons*) for these tapestries were often sketched by Parisian artists, prior to their being woven in a workshop in or near Brussels. In nature, orange blossoms show five-old symmetry and have five petals. Yet, such tapestries—months and years in the making, woven with the greatest of care—sometimes show orange flowers with only four petals, as in the case of one panel, *"La vie seigneuriale: La collation,"* at Musée de Cluny.

What role does the orange tree play in these tapestries? The answer lies in an examination of the "Dame à la Licorne" (Lady of the Unicorn) tapestries at Musée de Cluny. This set of tapestries has six separate panels. The first five panels are devoted to each of the five senses: touch, smell, taste, sight, and hearing. The sixth panel stands for intellectual pleasures.

What do those panels depict? In each, a Lady faces the Unicorn. The human figure and the mythical beast are shown within a lovely, flower-strewn garden. In some of the panels, other people and other animals can be seen as well.

I wish to focus on the garden itself, and more specifically on the orange tree. The utopian garden is shown as an island. Such an enclosed space is allegorical; only God can access its interiority. Moreover, the garden is circular, while a square pattern is superimposed by four trees at the four corners. Such circle-square superposition symbolizes the union of the microcosm—that is, man—with the whole of divine creation, the macrocosm.

And why four trees? The number four is reminiscent of the four rivers going through the Garden of Eden (as described in Genesis). It is also the number of the cardinal virtues: prudence, strength, temperance, and justice. It also stands for the four ways to read and interpret the Scriptures. The literal meaning coexists with the allegorical, the moral, and the anagogic.

Which trees are they? There is an oak tree, a pine tree, a holly tree, and an orange tree. The oak symbolizes vigor. In the allegorical interpretation, one has to break open the tough shell of the acorns to get to the edible kernels. The pine tree has a regal bearing and meaning. The holly stands for renewal and for the Resurrection.

And what about the orange tree? That it can bear flowers and fruit simultaneously (which is how it is displayed in the panels), appealing both to taste and to the intellect, speaks to the prodigious fertility of nature, and also to the generous profusion with which God lavishes His graces on us. The assimilation of the orange tree into the bountiful garden is a legacy from Isidore of Seville. He distinguished the garden from the wilderness (in his *Etymologies*) by the former being fecund whatever the season. Perhaps most important, because of the spherical shape and the color of the fruit, the orange tree is a solar symbol. This ties in with the solar nature of Christ ("I am the Light and the Life").

———

As documentary evidence, the tapestries woven in Flanders at the turn of the sixteenth century suggest that orange trees were exotic in Flemish culture during that period—witness the incorrect number of petals on the blossoms in some of them, obviously depicted from hearsay. A century later, citrus fruits proliferated in still life paintings produced and sold in the Low Countries. This shift deserves an explanation.

By the turn of the seventeenth century, the Dutch had become the richest people in the Western world. Their Protestant religion did not conflict with their making profits from rising capitalism. Their wealth came predominantly from international trade. The Dutch were the dominant naval power, and their ships brought goods from the four corners of the planet.

Affluent Dutch burghers could afford to commission and buy works of art. This enabled them to emulate what only aristocrats could do previously. Moreover, Calvinist values, with their emphasis on moral transparency, did not preclude flaunting one's material success. On the contrary, showing off was expected; it provided a model for others to imitate.

Thus, Dutch burghers hung still life paintings in their houses. These paintings mirrored the good life that the Dutch bourgeois were working so hard to attain: fancy foods, beautiful flowers, gorgeous rarities from nature. The underlying theme was unashamedly affluence and well-being. Still lifes depicted highly coveted but highly perishable gourmet delicacies, which in previous ages were the prerogative of the aristocracy: wild game, rare fish, spices, exotic fruits. No wonder that, during the seventeenth century, still lifes were produced at an all-time high rate, in the tens of thousands, on wood panels, canvas, or sheets of copper.

The ostensible subject matter could be flowers, exotic fruits such as citrus, other delicious foods and delicacies, fabrics imported from India and the Far East, or souvenir seashells from distant shores. These still lifes translate the cornucopia of antiquity into the pictorial language of the age of Vermeer. The style of these paintings was verisimilitude, a realism that did not shy from *trompe-l'œil*. Often in these paintings, a fly or another insect has landed upon an appetizing morsel, a piece of cheese or a mound of butter. The viewer is tempted to brush aside, or to pinch out and remove, the little intruder. Technically, this effect was assisted by the multiple layers of glaze applied by the artist. Lenslike, the glaze caught the light, which made for apparent depth and vivid colors. This was the recognized expertise of the Dutch painters: to represent meticulously, in the minutest detail, so that the colors gave the illusion of real, beautiful natural objects.

The Dutch still life is a precious snapshot of a moment in history, of the mental world—of the dream world, one might say—of the newly affluent bourgeois in the seventeenth- and eighteenth-century Netherlands. The seventeenth-century still lifes of Flanders and the Netherlands proper not only shared a style of painting, utter realism, and catered to the taste of the *nouveaux riches*, but their creators would seek clients in both regions of the Low Countries.

There was, for instance, Jan Davidsz de Heem (1605/6–1683/84), who divided his career between his native Utrecht and two decades in Antwerp. Why did he move to the major Flemish center of commerce? For what its markets had to offer. According to Sandrart, he moved to Antwerp because "there one could have rare fruit of all kinds, large plums, peaches, cherries, oranges, lemons, grapes, and others in finer condition and state of ripeness to draw from life."

De Heem needed such models because he specialized in mouthwatering pieces. His were lavish pictures of large accumulations of the most coveted foods, shown on tables covered with the rarest of Oriental rugs, bearing precious glassware, expensive decanters, and china and silver platters piled high with the most beautiful and exotic fruit.

The Dutch named these images of buffet displays *pronk stilleven*, or *pronk* still lifes. The adjective *pronk* was appropriate, as it conveys both ostentation and sumptuosity. Other artists also active in the genre were Abraham van Beyeren (1620/21–1690) and Willem Kalf (1622–1693). Citrus fruits did not appear only in *pronk* still lifes, however. One also finds them in other subgenres of the still life, such as the *ontbijt*: breakfast representations often featuring oranges or lemons. Willem Claesz Heda (1593/94–1680/82) made his name with such paintings.

Seventeenth-century Dutch still lifes show a frequent, recurring motif. A suitably ornate table is heavily laden with flowers, Venetian glass, porcelain, seashells, and seafood (such as lobster and oysters), plus an abundance of fruit. In the foreground of the composition, the artist often places a sample of his prowess. This feat of transcendental technical virtuosity is a peeled lemon, with the spiraling peel still attached to the unclad fruit, showing the delicate finery on the surface of the lemon and the lacelike lineaments of pith, in a regular geometric array, on the surface of the segments within.

You can see a lemon with its peel in Jan Davidsz de Heem's *Still Life with Fruit* (1652, Nardoni Gallery, Prague), in his *A Dessert* (1640, Musée du Louvre, Paris), in his *Still Life* (Kunstmuseum, St. Gallen), and in his *Still Life with Oysters, Lemons, Shrimps, and Fruit, with a Blue and White Earthenware Jug, on a Draped Table* (Otto Naumann, Ltd.); in his son Cornelis de Heem's (1631–1695) *Still Life with Fruit* (1665/70, Tokyo Fuji Art Museum); in Willem Kalf's *Still Life with Nautilus Goblet* (1660, Thyssen-Bornemisza Collection, Lugano) and *Still Life with Lemon, Oranges, and a Glass of Wine* (1663/4, Staatliche Kinsthalle, Karlsruhe); in Abraham van Beyeren's *Still Life with Fruit* (Pinakothek, Munich), *Banquet Still-Life* (Mauritshuis, The Hague), and *A Still-Life with a Nautilus Cup* (private collection); in Claesz Heda's *Breakfast with a Crab* (Hermitage, St. Petersburg); *Stilleven met vergulde bokaal* (1635, Stedelijk Museum, Amsterdam); and *Still Life* (1634, Boymans-van Beuningen Museum, Rotterdam); and in countless other works.

You are forgiven for skipping the previous paragraph! There is a plethora of such paintings. The sheer number of Dutch still lifes showing food in the seventeenth century boggles the imagination. They show mountains of food, and the artist renders it and the accompanying paraphernalia in baroque, not sparing the minutest detail.

One of the paintings just mentioned, *A Dessert*, at the Louvre, has distinct historical and art historical value. It was in the collection of the Sun King, Louis XIV, before 1683. And Henri Matisse paid close attention to it, first making a copy (1893), then basing an original painting on it (*Still Life after Jan Davidsz de Heem*, 1915, Museum of Modern Art, New

York). I shall return to the connection between Dutch seventeenth-century still lifes and Matisse's paintings of citrus.

―――――

Were depictions of goods such as luxury foods similarly appealing else-where outside the Low Countries at the same time? The answer is a qual-ified "yes." Similar pictures were painted in countries farther south. Con-trary to the stereotype of the Dutch, exuberance was the province of the Dutch still life. More restraint prevailed in Catholic countries such as Italy or Spain.

For instance, *Kitchen Still Life* (ca. 1640, Art Institute, Chicago), by the Italian painter Paolo Antonio Barbieri (1603–1649), shows a few mush-rooms, grapes, and a platter of almonds in the foreground, while the background holds a chestnut-filled basket and, on a stunningly white, unfolded piece of paper, a rather indistinct wooden object, a small vase perhaps. Likewise, *Still Life with Bowl of Citrons* (late 1640s, Getty Mu-seum, Los Angeles), by Giovanna Garzoni, is more in the tradition of botanical illustration.

Meanwhile, Spanish artists were painting still lifes in a style very much their own. With the same subject matter as their Dutch contemporaries, instead of emphasizing lavish accumulations, they focused on the beauty of individual life forms, on their contrasted hues and textures. By giving an intellectual and sometimes spiritual content to their work, rather than just basking in sensuousness, the Spaniards defined a new genre, known as *bodegones* (from *bodega*, a modest inn; the derisive name came from these paintings sometimes ending up as decoration in such places).

Bodegones depicted items in an enclosure. An atmosphere of serenity characterizes them. A *bodegón* would assemble, let's say, some fruit, a glass bowl or vase, an exotic bird, and the folds of a handsome fabric. Prior to a spiritual crisis during which he became a Carthusian monk, Juan Sanchez Cotán (1561–1627) brought *bodegones* to exquisite perfec-tion during the early baroque period (1600–1630). He imbued objects from everyday life—a quince or a cabbage—with palpable spirituality. Francisco de Zurbarán was another creator of *bodegones*.

Among other jewels, the Norton Simon Museum in Pasadena, Califor-nia, holds one of the treasures of world art, Zurbarán's *Lemons, Oranges, and a Rose*. It shows, left to right, a few lemons (citrons, to be exact) on a metal platter, oranges in a basket with a flowery branch on top, and on another, similar metallic platter, a rose and, next to it, a delicate cup half filled with water. These objects rest on a horizontal surface, truly more a stand than a table, of a dark, purplish color. The skins of the citrons and oranges beg to be touched, so convincing is the illusion.

This still life is enigmatic. Why show these precise objects? Why are they so assembled? As a *bodegón*, this painting is subverted by the presence of the rose, unusual in a depiction of food items. Moreover, the Zurbarán painting differs from many of the still lifes of the period in its starkness. The *trompe-l'œil* virtuosity only draws attention to the associated abstraction, to the unadorned and schematic stand holding the fruits, the flower, and the little cup.

Some believe that the only purpose of art history is to provide a reading of works of art and to spell out what they mean. Yet one has to resist the temptation to reduce this painting to its interpretation only. This caveat should be kept in mind when I divulge what the painting is about, beyond its ostensible subject matter.

Who painted this masterpiece? Francisco de Zurbarán (1598–1664) was the leading painter in Andalusia during the reign of Philip III and thus during the middle baroque period (1630–1670). A perfectionist, Zurbarán completed an estimated mere 250 paintings in his lifetime. The picture now in the Norton Simon collection was completed in 1633, possibly during a stay in Madrid, although Zurbarán lived and worked in Seville. Five paintings remain from his production that year, at the beginning of his career.

Zurbarán invested considerable work and forethought in this canvas. There exists a preparatory painting depicting the rose and the cup of water. In addition, Zurbarán had used the identical motif in two earlier, religious paintings. And radiographs have shown that the plate of citrons was originally flanked by a platter of candied sweet potatoes (*batata confitada*).

The Zurbarán painting is a technical gem, in its subtle rendering of perspective and in its playing off of the brightly lit (from the left) objects in the foreground against the dark background. Charles Sterling wrote of it, "The painter attunes the arabesque of a leaf to the curve of a lemon and makes the mouth of a cup respond to the opening of a basket."

A basket? Indeed, this canvas juxtaposes the crafted and the natural. Human products of metallurgy, basket weaving, and pottery hold products of nature. Is it mere coincidence that many—perhaps all—the objects depicted by Zurbarán share an exotic Asian origin? As already mentioned, citrus was brought to Europe from Asia during the occupation of Andalusia by the Arabs. Roses came from Persia. Porcelain was a precious material that the East India Companies of various countries

(Spain, Portugal, Britain, France, the Netherlands) were importing from China, as its English name reminds us.

Which suggests a conjecture: that the spirituality of this painting, which suffuses it, derives likewise from Eastern mysticisms, such as Sufi mysticism and its placement of the supreme value on purity (*safwa* in Arabic). Zurbarán has set up an illusion. With seemingly mundane subject matter, these representations of the everyday draw attention away from themselves. They are allegorical. To look at this painting is to open oneself, one's inner life, to a transcendental notion. What superficially appears as a *bodegón* is actually a religious painting.

Some authors see in it a symbolic homage to the Virgin Mary. This makes a great deal of sense. A key to this interpretation is the rose, which is not in water, but is juxtaposed with the water in the white porcelain cup: a flower is humbly presented, presumably to a lady. Thus, the horizontal surface can be seen as an altar, upon which various devotional offerings have been placed. Reinforcing this mystical allusion are various symbols associating divine love and purity (the rose and the water-filled cup) with chastity (the lemons and the oranges) and fecundity (the orange blossoms). In the preparatory work, the sweet potatoes likewise may be construed as an allusion to divine sweetness.

In the Zurbarán painting, one witnesses a key moment in cultural history. Its idiom is a legacy of the Muslim occupation of Spain, which brought with it the enduring concept of romantic love. As in the writings of the mystics Teresa of Avila (1515–1582) and Juan de la Cruz (1542–1591), both contemporaries of Zurbarán, sacred love is expressed in the language of profane love. In addition to Spain's Arabic legacy, the picture also makes at least one covert allusion to the Jewish component of Spanish culture: an important festival, Sukkot, in the Jewish religion requires citrons.

With the benefit of hindsight, the modern viewer can also see a reminder of the role of the Iberian peninsula not only as a melting pot in cultural history, melding the Jewish, the Arabic, and the Christian, but also in food history, as a way station in the westward movement of citrus fruits toward the Americas.

———

Representations of citrus in Western art followed a trend of gradually increasing familiarity. In the Low Countries in the early modern period, the trees were distant fantasies. In the Unicorn tapestries, they assume quasi-mythical status. With the advent of capitalism, the Dutch still lifes of the seventeenth century portrayed these still exotic fruits as valuables. Painters depicted them with the utmost attention to detail, as if they were jewelry or finery, which the preciousness of the fruits warranted.

Moving toward the nineteenth century, a downgrading is obvious as citrus fruits became commonplace in western Europe. Works of art document the change: why bother painting the mundane? Citrus fruits, which, except in a few production areas, were reserved for the upper classes at the time of the Renaissance, turned into a middle-class commodity. The Industrial Revolution was responsible not only for the increased means of the urban middle class, but also for the railroads, a major technological innovation that made it possible to bring citrus fruits to the bourgeois table. Accordingly, citrus fruits have only a fleeting presence in the art of western Europe until the 1870s. They appear only in still lifes, during the rare periods when that genre was back in fashion.

True, one still saw pictures of citrus in places remote from the production areas. They served the purpose of showing a desirable but unobtainable object of consumption. This was true from Oslo to St. Petersburg. To some extent, it was also true, at least in the early nineteenth century, in Boston and New York. In this respect, the pictures of citrus painted by Raphaelle Peale (1774–1825), who had his studio in Philadelphia, are noteworthy. His father, as well as other contemporary painters, looked down on the still life. They deemed the genre unworthy of professionals. Raphaelle Peale rebelled against this prejudice. He produced magnificent paintings, such as *A Dessert (Still Life with Lemons and Oranges)* (1814, National Gallery, Washington, DC).

Returning to the European continent, the Impressionists were responsible for the resurrection of citrus fruits as objects worthy of depiction. Paul Cézanne (1839–1906), of the legendary apples, would often include oranges in his still lifes. With him, the interest shifted to the light and the forms, away from the texture and the naturalistic details that the seventeenth-century Dutch painters had been so keen on. Vincent van Gogh (1853–1890), with his fascination with the color yellow—which some have blamed on absinthe and some on the professional disease of pica, which makes the sufferer crave camphor and turpentine—included lemons in his still lifes, such as *Still Life with Oranges, Lemons and Blue Gloves* (1899, Yale University, Paul Mellon bequest).

Following the Impressionists, Cubism had a predilection for the still life. Juan Gris (1887–1927) often painted citrus fruits, as in *Fruit Dish, Glass and Lemon* (1916, Phillips Collection, Washington, DC).

Henri Matisse (1869–1954) recapitulates Western painting from the seventeenth to the twentieth century with the theme of citrus as still life.

In 1893, as already mentioned, he copied the masterpiece by de Heem in the Louvre. He returned to that same model for his *Still Life after Jan Davidsz de Heem* (1915). But let me turn to another picture by Matisse, painted only a year later.

First, a description. *A Vase with Oranges* (now held in a private collection) was painted in 1916. Matisse shows these fruits as a treasure to be cherished. They stand out as the lone colored objects on the canvas, which is otherwise bathed in a cold, uniformly gray light. The oranges lie in a glass container in the shape of a chalice, as if held out in offering to God, or to the gods—in a surely deliberate wink to the Zurbarán painting. The light illuminating the scene is stark, but the oranges shine with an inner radiance. Strangely, they are much more lemon-colored than orange-colored. The viewpoint, not uncommon for Matisse, is from above.

In *A Vase with Oranges*, Matisse breaks with tradition. The subject of the title occupies the entire middle portion in the canvas, otherwise vacant of any feature but an indistinct gray wash. The viewer sees this bowl of fruit, on a table, from a standing position. And the framing is such as to truncate the whole top of the "window": we see only part of the scene, ruthlessly cut off at the top, where the edge of the canvas slices just above the two uppermost oranges. Likewise, we do not see the whole rim of the bowl; about one-third of the circumference is left out. This is a very deliberate excision and a partial exclusion of the viewer.

Moreover, a crescent-shaped segment of an orange lies abandoned, discreet and conspicuous at the same time, at the bottom of the stem of the vase holding the fruit, the vase dramatically foreshortened to emphasize the towering viewpoint. It nudges us to a dialectic of the outside and the inside, of what appears and what is real. The surface texture of the oranges, and their spherical shapes, are contrasted with the segment, a feature of their anatomy, as it were. The skin is shown together with the flesh, the whole together with the part—the amputated limb, so to speak. The date of the painting, 1916, reminds us of another amputation. The Great War (Matisse was too old to have been drawn in as a soldier) was in a stalemate of horror. And this painting is not an escapist fantasy, suffused as it is with the bleakness of the times.

Paintings since the Renaissance, as in the case of the Zurbarán painting, have aimed at showing the "view through the window": the eye of the artist looks at the world, while the hand, the wielder of the brush, proceeds to transfer the vision onto the painted panel or canvas. In such paintings, the subject matter dictates the positioning of the "window." A bouquet of flowers would be shown centered on the canvas and viewed frontally, the eye of the painter appearing (falsely) to be aligned with the object depicted. This used to be one of the implicit conventions in art. One can muster ample support for this convention from both optics and psychology: an eye examining anything is in constant motion, both to focus on the object of attention and to see it from all its visible sides.

Matisse was steeped in that tradition. He loved to duplicate an imaginary window, assumed by the laws of perspective, with a real window. Many of his paintings depict a room, whether in Collioure or in Tahiti, with a window or a door opening onto a seascape outside. In *A Vase with Oranges*, he did something completely different: he refused to allow his subject to dictate how it would be framed.

There are other indications of the new logic at work in this 1916 painting. The bowl of oranges has dropped from all-important subject to mere pretext. The painting is calling attention to itself. The subject of the composition is the composition itself. Matisse has left (or dropped) other clues as to the genuine object being depicted. The colors are deliberately wrong, the oranges are yellow and somewhat greenish in patches, not golden.

The painting by Zurbarán, as we saw, aimed at religious emotion. It succeeds in gorging itself on realist representation. In the obverse paradox, the Matisse painting of oranges jettisons traditional rules of representation. In so doing, it achieves a fullness of emotion.

This emotion came from Matisse's passion for oranges. The sight of them caused small daily epiphanies. Oranges were portents of joy, of the beauty in life. One of the proudest moments in Matisse's professional life was when Picasso in 1945 purchased his 1912 *Still Life with a Basket of Oranges*. This gave such pleasure to Matisse that henceforth, on New Year's Day, he would have a basket of oranges sent to his friend and great rival.

————

But why did Matisse travel such a route? Could he not have continued to paint in the classical realist style of Zurbarán, with which he was so familiar from his early copy of the de Heem picture? There were in his time, and there still are, painters who continue to steep themselves in this Beaux-Arts tradition. They produce commissioned portraits. University halls all over the world are hung with pictures of presidents and chancellors in that style. Such painters do photographic still lifes and landscapes in addition to portraits of their wealthy clients.

To ask the question is to answer it: traditional realist representation is a blind alley. One reason is obvious, even considering only the spirituality of Zurbarán's masterpiece. Our minds are thoroughly de-Christianized by comparison with the seventeenth century. And our jaded eyes may see this picture as other than an instance of realism only with an effort.

Facsimile depictions of persons (portraits) and things (still lifes, with bouquets of flowers, bowls of fruits, and so on) reached an acme during the age of Vermeer (1632–1675) and of Zurbarán (1598–1663). From that point in history, later artists discovered other avenues. As Heraclitus, the Greek philosopher of antiquity, wrote in one of his best-known surviving fragments, "One does not dip the foot twice in the same stream." Post-Vermeer art gradually underwent a split between repetitive painters, who kept rehashing the old formulas of so-called realism, and innovative artists, who sought and found other means of expression for both art and the depiction of reality.

The former organized themselves to fulfill the desires of the well-to-do bourgeois, who valued art foremost as a commodity. They became the successors of the artists of the Renaissance who painted for patrons, making a living from commercial art. They became known as the Academic painters, exemplified by the *pompiers* (uninspired artists) of the second half of the nineteenth century. The innovative artists, on the other hand, made a clean break with this tradition, starting with Manet. Such painters, the Impressionists in particular, took their cues from hints already present in the art of Vermeer (and of Zurbarán). Malraux has argued that Goya was the first of these revolutionaries.

The rest of the story, featuring the Fauves, Cubism, Surrealism, and the subsequent rise of abstraction, is so well known that it need not be repeated here. Today, the commercialization of bourgeois ideological values has become pervasive, and representational painting is very much in demand. The evidence for this is the constant vogue of the Pre-Raphaelites—all the way down to the most extreme, the commercialism of a Thomas Kinkade (1958–). The buzzwords of this subset of commercial painters of our time are tradition and craftsmanship—academic methods of drawing and painting.

———

However, commercial art can have both value and creativity, as I will now demonstrate by way of citrus crate labels. This original art form began with the shipping of citrus fruit from California to the rest of the United States. Florida followed suit. Between 1885 and 1955, wooden crates bore colorful labels with brand names, information on the place of origin, and the grade of the fruit; there were subvarieties for lemons, oranges, grapefruit, and so forth. These printed paper vignettes have become collectors' items. There is a whole cottage industry, on the Internet in particular, trading these artifacts. They are wonderful pieces of Americana—wonderful because many are artistically valuable, besides providing the social historian with interesting documentation.

———

First, an outline of the history of citrus labels. As soon as the transcontinental railroads reached Los Angeles and growers started shipping oranges to markets in the eastern United States, Southern California farmers became intent on cashing in on the new bonanza. One might call it a second gold rush. It contributed significantly to the influx of people to California during the 1880s, when the state's population increased by some 345,000.

To producers, labeling the crates of oranges they were shipping across the country was a matter of pride. It was also a means to advertise California and the California dream through their produce. It was not only oranges that were being sold, but also contact with a distant wonderland. Later on, Hollywood would exploit a similar vein of magic with its movies.

From the start, citrus labels drew on existing expertise. By the 1870s, the labeling of wine bottles was nurturing an active business among the fifteen major print shops in San Francisco, which used lithography as the technique for reproduction. Max Schmidt, a German immigrant, substituted zincography in the 1880s, a process using metallic plates instead of wood or stone engraving. Hence, the shops' experience in designing wine labels was easily transferable to the printing of inexpensive and colorful citrus crate labels.

Meanwhile, the California citrus growers had organized themselves into a cooperative. The Southern California Fruit Exchange, later renamed the California Fruit Growers Exchange, standardized crate labels in size (a 10″ × 11″ paper rectangle for oranges, a little less for the smaller lemon boxes), in the technical information they carried, and in the recommended mention of the Sunkist trade name. Otherwise, in terms of the design of the individual label, diversity was the rule.

In spite of that diversity, an overarching style prevailed. Artworks belong to a given style. What is true of the art shown in galleries and museums is even truer of commercial art. Citrus labels thus belong to one of three successive periods, each with a distinctive style.

From 1885 to 1920, the style was naturalism. It took its edge and its emphasis from American popular art of the time. These labels, postcard-like, depict landscapes from California, typically a citrus grove set against a background of high mountains. Besides such scenery, California plants and other wildlife are also often shown. Soft, muted colors prevail. The citrus labels were to appeal, it was thought, to the housewife buying the fruit in the market.

However, a study commissioned in 1918 by the California Fruit Growers Exchange led to a changed perception. The jobber at auction was the person to be targeted, and as a rule, this person was a man. The labels were accordingly made considerably more macho, with themes of outdoor adventure and of California as a frontier.

This led to the starker, more brutal and modernistic label style that dominated the 1920–1935 period: clear, simple images, overrun with powerful titles and slogans. The label had to be identifiable on sight, from a distance, and easily remembered. It was essential to commercial success that the wholesale dealer, purchasing fruit at auction, be browbeaten into brand loyalty. Arguably, the year 1930, at the start of the Depression, was the zenith year for citrus labels. California growers were then shipping about 2,000 distinct brands.

The third and last period, 1935–1955, saw the demise of the citrus crate label as an art form as a consequence of World War II. Metal litho plates were scrapped in the war effort. Moreover, labor-intensive wooden crates were jettisoned in favor of preprinted cardboard boxes—and the colorful labels were no longer called for. During this last period, the dominant style became that of a commercial logo: often showing an orange, together with a forceful inscription in large, three-dimensional lettering, made monumental-looking with airbrushed drop shadows, sometimes slanted across the label, sometimes along a curve.

All in all, 8,000 distinct designs are estimated to have adorned more than 2 billion orange crates over a 70-year period.

The story in Florida is somewhat different. All labels were registered with the State Department of Agriculture. There was a color code: blue for Grade A, red for Grade B, and green for Grade C. The major print shop was the Florida Grower Press in Tampa. About 1,000 9″ × 9″ citrus crate labels from Florida are known today.

How did citrus labels achieve their main aim, to be memorable? Earlier in the tradition, they reflected the glow of the California dream and depicted local natural wonders. For instance, Valencia oranges from the Limoneira company in Ventura County, started in 1893, were given the brand name "Bridal Veil." The label showed the fruit, leaves, and blossoms next to a picture of the famous waterfall at Yosemite. Charles C. Chapman, a grower in Fullerton, shipped his late Valencias under the name "Golden Eagle." An image of the awesome bird with its wings outstretched, soaring above a bunch of oranges, combined the two mystiques of California and of America, of gold and of the eagle.

Sometimes, seemingly incongruous symbols created powerful images with a definite surrealistic appeal. I think of the "Mazuma" label—a nice memorable name that rhymes with Montezuma—with a half-peeled orange, as if hand-held in its regularly incised and partly removed skin, above what may be a topographic map, and next to a bag of gold coins ("mazuma" is a slang term for money).

At other times, the label is made even more memorable by an irrational, rather than by a readily understandable, incongruity. This is the case with "Green Rings," a brand name used by the McDermont company in Riverside. It shows the fruit sitting above four hollowed circular green slabs of an unidentifiable material. Clearly, the designer meant to leave an indelible image in the mind of the viewer, made more so by its total arbitrariness.

––––––––

Repetition was a means to instill brand loyalty in the jobber. The crate labels of the middle period are heavily self-referential. As mentioned earlier, a picture of an orange is usually present, even though the buyer at an auction knew that he was purchasing crates of oranges. Such redundancies could not but be deliberate: the "Caledonia" brand, from the Placentia Mutual Orange Association, in Orange County; "Upland Gold," from Upland Groves, in Upland, California; likewise, "Belle of Piru," from the Piru Citrus Association, in Piru, Ventura County. The self-evident purpose was to imprint the repeated word (Orange, Upland, or Piru) in the mind of the jobber.

But in my opinion, the prize goes to the "Index" brand label: a finger of that name points to a crate of oranges, most of which are shown in their paper wrappers. The short side of the box also bears the label so that the self-reference is reiterated. The long side of the box has the words "Index brand" in between two hands with the index fingers pointing at it, quotation mark–like. The bottom plank has the address, La Habra Valley, and, again, the contents, "Lemons–Oranges."

––––––––

But let us return to the first of the three styles, whose labels were likely to show scenes from Californian landscapes. My favorite is "Home of Ramona," from the Piru Citrus Association in Ventura County, copyrighted in 1900 by Elpiano del Valle. It advertises Valencia oranges, one of which, marked Sunkist, is shown in its paper wrapper in the lower left corner (which dates this particular label as post-1914). There is a house in the foreground, in Spanish California style, with closed shutters, a covered porch for an entrance, columns, and flowering shrubs. The orange grove is shown in the middle ground. And the background is occupied by tall snowy mountains—actually much more impressive in the illustration than are the actual mountains one sees in Ventura County! The inspiration for the label was Helen Hunt Jackson's novel *Ramona*, set at Rancho Camulos.

Such a label is similar to a label on a French bottle of wine depicting a *château*; that is to say, the wine's locale of origin. One might argue that these citrus labels are borrowed from those on French wine bottles—a complete imitation, down to the image of the owner's house and the surrounding scenery.

But what works for a French wine in France need not succeed in the United States. In the French culture, the admittedly untranslatable notion of a *terroir*—a grassroots area or a territory comes closest to a translation—is crucial to the identity of people and of products both. De Gaulle once made a quip about the impossibility of governing a country with no fewer than 350 different cheeses! Each such cheese comes from a different place. Its flavor and perfume reflect the local geography, climate, soil, historical tradition, and so on.

The French word *terroir* carries thus the combined notion of grassroots, tradition, and conservatism, and also of excellence. The whole system of *appellation contrôlée* for French wines is a means of associating a brand name with the authentic *terroir* the wine has been nurtured in. And the label on the bottle, with the image of a *château*, is a visible icon of such baggage.

Little did the grape and citrus growers in California, who started by imitating such wine labels from Europe, suspect how much of a cultural gap their labels attempted to span. This could not last. The notion of a *terroir* is too foreign to Americans, for whom a more important notion is mobility, the freedom to move about as is their wont. Americans strike Europeans as a people on the move. They often equate social success with geographic mobility. To stay in one place for more than a few years, or worse, for one's whole life, can be synonymous with stagnation.

Hence, the static postcard-like labels were replaced, starting after World War I, with images of American mobility and speed. Many labels featured freight trains rushing the citrus crates toward the consumers. The "Double A" is one such example. In the 1920s, with the development of commercial aviation, one also starts seeing pictures of airplanes. "City in the Sky" is a picture of a futuristic city floating in the clouds, anticipating *Star Trek*, complete with an airship. "Airship" is indeed another brand name, with an airplane for a logo. Likewise, one also finds a picture of a glider, for the brand "Glider."

More generally yet, citrus crate label art shows us Americanization at work. A useful parallel would be the movies, or better yet, animated cartoons, whose elevation by Walt Disney and others into a genuine American art form ran parallel to crate label art, and during the same period.

And how is such Americanization effected? The very first insight into anything American is the immensity of the continent and the need for countervailing unifying forces. A second feature is immigration, since the United States is a country of recent immigrants.

Accordingly, identity is rooted in history turned into mythology— such as the taming of the Wild West. The same factors that gave rise to the Western as a movie genre influenced citrus crate labels. Thus, Native American and Western scenes and Native American warriors often appear on labels: in "Bronco," a mounted cowboy gallops on sand. "Indian Belle" and "Indian Brave" show the courting of a maiden sitting on a rock. "Kaweah" shows another young Native American woman; "Pala Brave" displays a Native American in a headdress. "Rocky Hill" shows a Native American chief with his horse on a hill; and so on.

The imagery on these crate labels is, in short, very much the stuff taught to young American children in elementary school: American heroes such as George Washington and Abraham Lincoln; the beauty of the unspoiled virgin land; the reenactment by Americans of the age of chivalry with knights and ladies that is part of the ideological makeup of the Old South, *Gone with the Wind*-style; and the battle for the conquest of the West against noble and fierce Native American warriors.

Label art is thus generally very American in its subject matter. It seldom depicts exotic scenes. There is one exception, however. A subtheme of crate label art is, somewhat surprisingly, Scottish. Several labels show tartans of various kinds. One beautiful picture, for the "Caledonia" brand (Placentia, Orange County), shows a thistle on a background of tartan cloth. "Tweed" likewise shows a piece of such woven fabric. The Strathmore Packing House used several labels with Scottish subject matter: in "Scotch Lassie Jean," an attractive girl in traditional Scottish costume smiles and waves; in "Strathmore," a bagpipe player, also in full Scottish garb, marches on, with a thistle again in the background. There is "Kiltie" too, and many others.

The explanation for the association of citrus with Scotland and things Scottish is the supposedly Scottish origin of marmalade. Marmalade and toast are an integral part of many an American breakfast menu. Since the California Fruit Growers Exchange started heavy promotion, in the 1920s, of a glass of orange juice at breakfast, any connection with breakfast was welcome in order to sell more citrus. Indeed, one of the most beautiful vignettes is "Tartan," from the Corona Foothill Lemon Company (in Riverside County), which entices customers with a grapefruit on the breakfast table.

––––––

Emblematic analysis is a powerful means for better understanding advertising. An emblem—collections of emblems were first published in book form by Andrea Alciato, a law professor at University of Pavia, Italy, during the Renaissance—has three parts. An enigmatic image, *pictura* in Latin, is accompanied by two textual elements: a motto or caption, *inscriptio*: and an explaining commentary, the *suscriptio*.

It is my contention that any advertisement still conforms to the model first introduced in the Renaissance. Take as an example the handsome "Airship" label. Fillmore Citrus devised it at the start and remained loyal to it. Initially, this image showed a dirigible hovering over a citrus grove. It then gradually evolved, adapting to progress in the aircraft industry.

In the label I am looking at, the image (*pictura*) shows a four-propeller airplane, a passenger plane (as we can tell from the clearly visible windows) descending toward the ground. Upon checking, this is a purely imaginary aircraft, unrelated to an actual airplane of the time. In the foreground, at bottom left, is a wrapped orange, clearly bearing the Sunkist imprint—a characteristic of all Fillmore Citrus labels. The image is doubly puzzling: why is the airplane headed down? And why is it bathed, from underneath, in a golden glow illuminating the bottom of the wing and the horizontal stabilizer in the back? The answer to the second question is obvious: the gilding is imparted by the Sunkist orange. As to the first question, the implicit message is that the plane is about to make a landing in the El Dorado at Fillmore, because the passengers are clamoring to get out and to sample the golden fruits.

THIRTEEN

Preserving Nature—or Changing It?

La naranja, por la mañana, sana, al mediodia, pesa, por la noche, mata.
(An orange, in the morning, healthy, at noon, heavy, at night, 'tis a killer.)
SPANISH PROVERB

Do we save artifacts out of concern for our own mortality? A dried flower within the pages of a book. Egyptian mummies. The enigmatic portraits from the Fayoum. Animals, former pets or hunting trophies, stuffed by a taxidermist. These objects attest to a will to preserve traces from once-living organisms. One could make a case that all culture fulfils a similar need.

Yet more interesting, why does the brain seek return to earlier states of consciousness? Has memory some survival value for the individual? For humans as a group?

Proust's magnum opus, *A la recherche du temps perdu*, bravely explores the mysterious territory of remembrance. He gives attention to olfactory and gustatory memories, his recollection of the taste of the *madeleine* being a justly famous episode. Such memories are crucial to our appreciation of food and wine. Not unlike a classical music lover listening to a symphony, craving the sound of, say, the clarinet while also relishing its merging with the other instruments in the full sound of the entire orchestra, a gourmet may take pleasure in discerning a single note—such as the fragrance of violets in a bottle of Bourgueil—or in savoring the whole complex bouquet of flavors and smells that a well-prepared dish provides.

175

As it turns out, citrus fruits exert a powerful hold upon sensory memories. After presenting some of the empirical evidence in support of that claim, this chapter will analyze some of the culinary tricks for recapturing the taste of a tangerine, the tartness of a lemon, or even the odor of a grapefruit warmed by the caress of the sun.

Were these ploys devised, like perfumes, because citrus was a luxury for centuries? Probably. The evidence is rich: my selection will include citrus-flavored drinks and candy as well as dried citrus peel, whole and powdered.

Many commercial drinks, most of them carbonated, are citrus flavored. This is the preferred taste in soft drinks. Traditionally, it is imparted by essential oils extracted from citrus peel, which contain the two limonenes, L-limonene—which evokes the orange—and/or D-limonene—which evokes the lemon. Alternatively, more cheaply but not equivalently, the multicomponent essential oil is replaced with a single synthetic chemical.

The traditional method of flavoring soft drinks, using an essential oil, presents a technical difficulty: oil does not mix with water. An emulsifying agent is needed. In very small amounts, it turns the oil into microdroplets suspended in the water medium, so tiny that drinkers do not even suspect their presence. And, of course, such an emulsion must withstand carbonation without breakdown.

From soft drinks, let us turn to candy. This substance is among the richest and the most diverse in food history. To sample the kinds of sweets that Europeans enjoyed prior to the early modern period, try a *calisson*, a honey-based specialty from Aix-en-Provence, a cousin to *nougat* and to the Spanish *turron*, which go back to Arabic occupation and al-Andalus. Candy of various kinds reached western Europe from the Middle East through Spain, Portugal, and then Italy. Candy to this day often retains ingredients used during the late Middle Ages and the Renaissance. They smack of the inventory of goods in a Venetian galley: gum arabica, gum acacia, gum tragacanth, saffron, and so on.

––––––––

When cane sugar reached Europe in large amounts, beginning in the sixteenth and seventeenth centuries, it replaced honey as a sweetener. Myopic food historians date candy to the introduction of beet sugar to western Europe in the early nineteenth century. If, indeed, manufactured confections originated at that time, the history of sweets is definitely longer.

How were lemon drops made, for instance? Their gelled texture depended on pectin from citrus peel and apple pomace. They were flavored with lemon oil, and they were colored yellow with a little saffron. Only toward the end of the nineteenth century did artificial coloring start replacing natural dyes in candy. Orange drops were prepared in like manner, with oil of orange or oil of neroli supplying the flavor.

The challenge was, and remains, simulating in candy the sweet and sour combination so characteristic of citrus fruit. Candy is essentially made of sugars, such as glucose, fructose, maltose, or sucrose, and sugar-related chemicals, such as mannitol or sorbitol, so, most of these molecules being sweet (very few sugars are actually sweet-tasting), the first part of the equation is straightforward. With citrus flavors, the second part is easy too. The well-named citric acid is the usual ingredient imparting sourness to candy, whatever the desired hardness, crystallinity, or chewiness. True, other organic acids outdo citric acid as food ingredients. Malic acid, thus named because it was originally isolated from apples, is superior to citric acid as a flavor enhancer and for its persistent sourness. Tartaric acid, a wine by-product, is intermediate in sourness between citric and malic acids.

———

Now to dried citrus peel. Many an Alpine home, whether a chalet in the Valais, a small farm in Val d'Aosta, or an old stone house in Haute-Savoie, harbors a modest ornament on the wall of the kitchen or the living room— a garland made from dried fruit. It may have been the handiwork of a woman in the family, or it may have been done by one of the children as an elementary school project.

The elements of the wreath eventually become hardened. They may be a little dusty, too, if the garland has not been replaced yearly in a holiday ritual whose deep meaning is fertility, and whose intent is to preserve the memory of a warm summer throughout the long, bleak winter.

The sunny emblem I hold in my hands while writing this paragraph reminds me of my childhood in the French Alps. It is strung on a piece of raffia. A half-dozen larger and roughly spherical elements are separated on the string by alternating lemon and orange slices. The lemon slices are thicker and have turned a creamy white. The thinner orange slices have darkened into a deep, ruby or rusty orange. What exactly were those larger items originally? Apples? Tomatoes? Clementines? They have shriveled with age. Their smooth, polished skin has the lacquered and timeless sheen of an antique violin.

A mere decoration? I have seen these garlands used, for example, on a cold winter night. A slice is placed in a hot cup of black or herb tea, adorning and perfuming it. The practical and the symbolic thus merge. Just like the human spirit, summer endures through winter. The memory of summer sustains one, holds the spirit aloft through the harsh season.

The garland just described is a memory preserver: the dried fruit retains some of the original flavor and aroma. To some extent these become concentrated in the process.

Prior soaking in a chemical and then drying kills germs, thus preserving the flesh or fiber from decay. Mummification takes advantage of this tendency, as does the salting of foodstuffs such as dried meat, codfish, dry sausages, cheeses. From time immemorial, humankind has protected its protein supply through salting.

In like manner, candying combines infusion of a fruit with sugar with drying. Since at least the Renaissance, if not much earlier, European gourmets able to afford it have feasted on candied fruit. By and large, confectioners in western European cities such as Rome, Vienna, Frankfurt, Liège, Paris, and Utrecht were thus acquainting their customers with otherwise unattainable fruits from Asia, such as dates or figs, among which citrus fruits of all varieties were the most abundant.

Furthermore, candying is an easily implemented procedure, with the rather amazing result of substituting a sweet for a sour taste. Even if the original fruit tastes rather tart, a candied orange has become all sweetness while retaining (or so it seems) the wholesome taste of an orange. The unmistakable orange (or lemon, or tangerine, or clementine) aromas have been preserved.

When one bites into a segment of fresh orange, the skin is punctured, and the fruit bursts into a juicy mouthful, combining acidic and sugary tastes. With a candied orange, all the tartness is gone. One has a similar impression in eating other dried fruits, whether apples or raisins: the sweetness has just been heightened, reinforced. The truth, of course, is different. During candying, the fruit has lost all the juice from its cells. Their contents have been fully replaced with sugars.

———————

Dawson, the writer of a cookbook most influential during the sixteenth century, gives a recipe for candying oranges or citrons:

Take Cytrons and cut them in peeces, taking out of them the juice or substance, then boyle them in freshe water halfe an hower untill they be tender, and when you take them out, cast them into cold water, leave them there a good while, then set them on the fire againe in other freshe water, doo but heate a little with a small fire, for it not seeth, but let it simper a little, continue thus eight daies together heating them every day inn hot water: some heat the watre but one day, to the end that the citron be not too tender, but change the freshe water at night to take out the bitternesse of the pilles, the which being taken away, you must take suger or Honey clarified wherein you must the citrons put, having first wel dried them from the water, & in winter you must keep them from the frost, & in the Sommer you shal leave them there all night, and a day and a night in Honie, then boile the Honie or Sugar by it selfe without the orenges or Citrons by the space of halfe an hower or lesse with a little fire, and being cold set it againe to the fire with the space of halfe an hower or lesser with a litle fire, and being colde set it againe to the fire with the Citrons, continuing so two mornings: if you will put Honnie in water and not suger, you must clarifie it two times, and straine it through a strayner: having thus warmed and clarfied it you shall straine it and sett it againe to the fire, with citrons onely, making them to boyle with a soft fire the space of a quarter of an houre, then take it from the fire & let it rest at every time you do it, a day & a night: the next morning you shall boyle it again together the space of half an hower, and doo so two morninges, to the end that the Honie or Suger may be well incorporated with the Citrons.

Thus at the end of the fifteenth century, honey and sugar were used as sweeteners. Note the time required for Dawson's recipe. Today, one may proceed more rapidly in a modern version:

CANDIED CITRUS STRIPS

2 or 3 citrus fruits, brightly colored, but not limes
3 cups granulated sugar

- Remove the peels from the fruit and cut them into 1/4 inch strips.
- Drop the strips into boiling water and allow them to simmer for 2 minutes.
- Refresh the strips under cold running water in a colander and repeat the boiling process once more, using fresh water.
- Dissolve 2 cups of the sugar in 1 cup water and slowly bring to a boil.
- Add the blanched citrus strips and simmer for 20 minutes or until they become translucent.
- Cool the strips in the syrup, then dry on a rack until slightly tacky, about 2 hours or so.
- Roll the strips in the reserved cup of sugar and store them in an airtight container.

Yes, this is most delicious. It is as complete a denaturing, though, of the fragile molecules ensconced in the peel that one could achieve. Simmering in hot water for about an hour will see to alteration of the molecules, if not to their destruction.

Let us not exaggerate. During drying and/or candying, many chemicals are retained. Saved from destruction, they offer a lingering trace of aroma.

A better idea, if one wishes to retain a whiff of the natural aroma, is to take a leaf from the album of chemists. When they isolate a substance from a plant, they start by extracting it with a solvent. The procedure draws on the solubility of various chemicals in various solvents, which a chemical laboratory holds in diverse supply: water, alcohol, ether, petroleum ether, benzene, and other hydrocarbons distilled from petroleum, such as hexane. The goal—retrieval of a natural aroma—is elusive for two reasons. Some of the ingredients will dissolve in one solvent—say, water—but not in another—say, alcohol. And some are chemically changed by the extraction process itself. Nevertheless, from the twelfth century onward, once Europeans had learned (from the Arabs originally) how to purify alcohol by distillation, alcoholic extraction became a cottage industry. It provided remedies (faith healing) and it also produced the ancestors of liqueurs such as Bénédictine, Cointreau, and Chartreuse. Needless to say, alcoholic extraction was applied to citrus peel: witness orange wine.

ORANGE WINE

Peel from a half-dozen oranges
1 liter fruit alcohol, at least 90 proof
2 lbs granulated white sugar
3 bottles dry white wine
2 bottles sweet white wine
A half-dozen peppercorns
A half-dozen cloves
A quarter of a cinnamon stick

- Remove the white pith, as much as possible, from the orange peel.
- Put the peel in the fruit alcohol, together with the spices.
- Let the preparation marinate for 3 weeks.
- Stir in the sugar with a wooden spoon until dissolved.
- Add the wine.
- Allow the mix to marinate for another 3 weeks.
- Filter out all of the solids.

———

Le cru et le cuit (The Raw and the Cooked) is the title of a book by the anthropologist Claude Lévi-Strauss, based on tales by Amerindians and suggesting structuralist interpretations. As we saw with candying, simmering destroys some of the molecules in citrus peel. Most such terpenes cannot withstand an elevation of temperature without some structural damage.

However, cooking is not uniformly inimical to aromas. Common experience tells us otherwise. Delightful fragrances emanate from a pan on the stove. The kitchen is the location not only for the destruction of natural aromas by cooking, but also for the appearance of new aromas that cooking itself creates.

Citrus fruit obeys that rule. Cooking sets off new volatile molecules, some of which delight the nose. This is due to the chemical processes of caramelization and the Maillard reaction.

Caramel formation first: application of heat, in a first step, cleaves sucrose molecules present in citrus fruit into glucose and fructose. Once separated, these molecules, in a second step, combine when heated into long polymeric chains of caramel.

As for the Maillard reaction, discovered at the turn of the twentieth century, it is the main producer of those appetizing scents from the kitchen. The combination of amino acids from food proteins with sugars when the two are heated together results in volatile molecules, which we like to think smell heavenly. The Maillard reaction explains why meat and vegetables turn brown when cooked, releasing an aroma in the process.

A favorable circumstance for both caramelization and the Maillard reaction is the heating of a sugar syrup. A sugar syrup, of course, is any aqueous solution containing a high concentration of sugars—and a tiny amount of proteins and amino acids, nevertheless more than enough for the Maillard reaction to take place. A chef may prepare a syrup by dissolving sugar in water—and water can absorb an awful lot of sugar. A syrup also results whenever a sweet juice, with dissolved sugars, such as orange or grapefruit juice, is subjected to heating. Losing most of its water to evaporation, it gradually becomes viscous. Hence, heating, or better yet, simmering citrus juice or, for that matter, chunks of whole citrus fruit combines nature and culture, two categories that Lévi-Strauss has invoked repeatedly.

I have stored in a notebook the following recipe for marmalade, given me in Auckland by the wife of a colleague in the chemistry department of the university there, the best in New Zealand:

MARMALADE

As for the amount of sugar in this recipe, a good rule of thumb to follow is 3/4 cup of sugar for each cup of cooked fruit. If Seville oranges cannot be found, you might wish to add a couple of lemons to impart additional tartness.

2 large grapefruits
6 oranges, preferably Seville
1 cup white granular sugar, more or less
1 cup brown sugar

- Wash the fruit, slice thinly, and cover with water in a saucepan.
- Allow the fruit to stand, covered, overnight.
- The next day, bring the fruit and water to a boil and simmer until the fruit is tender, about 40 minutes.
- When the fruit is nearly tender, place the sugars on a sheet pan in a warm 275°F oven. Heat for about 20 minutes.
- Add the heated sugars to the fruit and continue to cook until the fruit sets. Keep a chilled plate in the freezer to test the thickness of the fruit throughout the final cooking process. Place a small amount of the cooking fruit on the cold plate. If it is done, it will set up quickly.
- Cool slightly, place in sterile jars, and process accordingly.

———

Marmalade, to me, brings back a sunny memory indelibly associated with an olfactory feast. Cedar Key, on the Florida coast, attracts tourists with a hint of wilderness and with large numbers of pelicans, graceful in flight. Humans have but a toehold there; the village is regularly swept by hurricanes traveling north from the Caribbean.

On a drive from Coral Gables, near Miami, my wife and I stopped at Cedar Key. With the luck of the road, we stayed in quite a nice motel. It did not advertise, but offered elegant suites. In the morning, we went out and purchased citrus fruit, the fully equipped kitchen giving us the impulse. We made marmalade, using the above recipe. During the remainder of our stay in Cedar Key, the apartment had a heavenly smell, somehow consonant with the aroma evoked by the name, Cedar Key. We bought a few empty jars at the same tiny local market where we had purchased the fruit and filled them with our production. And we drove off to the north, along U.S. 1, ultimately to New Hampshire, the marmalade nourishing us all the way.

———

To look up a word can be great fun. Some fancy anecdotal etymologies have been proposed for "marmalade." A few strive to confirm the Scottish origin of the recipe, which is probably a myth.

Some ascribe marmalade to Mary, Queen of Scots, who was briefly married to the Dauphin (the eldest son of the king of France). When she fell sick, "*Marie est malade*" was the French phrase used around her retinue, which gave rise, in this unbelievable story, to "marmalade." What is the connection with marmalade? Was she sick and wanted to eat orange jam? In a variant of such a royal episode, the queen was seasick, and *mer malade*, in French again, led to "marmalade!"

It is easier to believe in peninsular origins for the English word "marmalade," since the dish was made originally, in the seventeenth and eighteenth centuries, from Seville oranges exported to France or Britain. Indeed, the Portuguese language contains the word *marmelo*, for the quince fruit, which came, in turn, from the Latin word *melimelum*, used for certain types of sweet apples. That "marmalade" came from the romance languages of southern Europe is further confirmed by the early existence in Spanish of the word *mermelada*. The first occurrences in French go back to 1573, as *mermelade*, and to 1642, with the spelling *marmellade* (both indeed close to the Portuguese *marmelo*).

The English word obviously derives from *marmelada*, the name in Portuguese for quince cheese. Hence, the scientific question is that of transmission, direct or indirect, through the French. Direct transmission is implied by legend: when Catarina of Bragança married Charles II in 1662, she waxed nostalgic for quince preserves from home, as the story goes. However, "marmalade" was recorded in the English language as early as 1480, long before its earliest occurrence in French.

Occurrences of the word in English, under various spellings borrowed from the French, such as "murmblade," "marmalet," "marmalad," "marmaled," date back to the sixteenth and seventeenth centuries. Most often, they refer to a preserve made by boiling quinces.

But how did the word for "quince" come to mean the product of cooking citrus fruit with sugar? To answer that question, one needs to return to the Latin term, *melimelum*, for a honey-apple or a sweet apple. It came

183

in turn from the Greek *melimêlon*, a word meaning orange preserves or "marmalade."

The confusion between the quince fruit and marmalade goes back to ancient Latin and Greek and to a major lexical distinction, between two kinds of fruits—nuts (*nux* in Latin) and those having pits or pips, which the Latin language endowed with the generic name *malum* (derived from the Greek *mêlon*). An apple was thus known to the Romans as *malus*. And applesauce came to be known in Latin by the derived word *melata*. Hence, since Greek already named *melimêlon* what we know as a marmalade, Romans spoke by analogy of *melimelata* for an orange marmalade. As for the Portuguese *marmelo*, it came from the same Greek word *melimêlon*, but with the meaning of a honey-apple; that is, of a quince having been grafted onto an apple tree. Our modern melons also go back to the same Greek and Latin names for soft fruits with seeds inside them.

Is there a relationship between "marmalade" and *meli-melo*, which in Italian and French means a mish-mash, a disordered mix of various things? Today, many restaurants will carry on the dessert menu a "fruit melimelo," meaning a fruit salad of sorts.

———

Chutney is a condiment that the British discovered during their colonial occupation of India and shared with the world. A mish-mash indeed, it is a strong, hot relish, according to the *Oxford English Dictionary*, "compounded of ripe fruits, acids, or sour herbs, and flavoured with chillies, spices, etc." In other words, a chutney combines three taste types, sweet, sour, and spicy. There are numerous kinds of chutneys. Many can be made from citrus, as in the following recipe:

LIME CHUTNEY

12 limes, halved
1 medium onion, peeled and quartered
4 hot green chili peppers
1 inch ginger root
4 oz seedless raisins
7 green cardamom pods
1 tablespoon black peppercorns
1 tablespoon coriander seeds
1 tablespoon mustard seeds
4 dried red chili peppers

1 1/2 cups cider vinegar
3 tablespoons coarse salt
1 lb light brown sugar

- Juice the limes. Discard 6 lime halves.
- In a food processor, combine remaining 18 lime halves, green chili peppers, onion, ginger, and raisins. Chop finely. Place mixture in a non-metal bowl.
- Open cardamom pods. In a heavy skillet, toast peppercorns, cardamom seeds, mustard seeds, coriander seeds, and the dried red chilis for about 3 minutes, stirring constantly. Let the spices cool on a dry plate, then grind finely.
- Add spices, lime juice, sugar, and vinegar to the chopped fruit mixture. Stir thoroughly, cover, and let steep at room temperature for 2 days.
- On the third day, pour mixture into an enameled pot (no stainless steel!), add salt, and bring to a boil slowly. Simmer, uncovered, for 30 minutes.
- Place in prepared clean jars. Close jars with a tight-fitting lid. Store in a cool place.

———

Chutney is one of the uses suggested for one of the latest citrus fruits to reach our shores, whose marketing is still in the planning stages as I write: the desert lime, *Citrus glauca* or *Eremocitrus glauca*, depending on the authors. Found in the Australian wilderness, it grows in semiarid areas of New South Wales and Queensland, on a broad range of soil types. Desert lime has the shortest flowering to harvest period of any known citrus, with a midsummer harvest following the spring flowering. The fruit is yellow green, tiny (just two centimeters in diameter), often seedless, and juicy, with the taste of a tart West Indian lime.

———

A treat akin to marmalade is lemon (or orange) tart. The mother's traditional role in many a culture has been to provide warm affection, often in the tangible form of sweets and desserts. I am fond of the expression "as American as Mom and apple pie." There are French equivalents of apple pie; they include *crème caramel*, *île flottante* (otherwise known as *œufs à la neige*), and *tarte au citron*:

TARTE AU CITRON

Tart shell:
1 1/4cup all-purpose flour

185

1 stick (1/2 cup) cold unsalted butter
1/2 teaspoon sugar
1/2 teaspoon salt
2 or 3 tablespoons ice-cold water
1 egg yolk

- Preheat oven to 350°F.
- Combine flour, salt, sugar, and butter in a food processor and blend until coarse crumbs form.
- Turn the mixture into a bowl. Sprinkle ice-cold water over it and mix with a fork until it forms a ball. Add additional drops of water if it is too dry.
- Form dough into a disk, wrap in plastic, and refrigerate to rest for at least 30 minutes.
- Roll out dough on a lightly floured board to fit a 9-inch tart pan.
- Fit dough into pan and prick all over with a fork.
- Line the tart with aluminum foil and fill with pie weights (or with rice, beans, etc.)
- Bake shell until slightly firm around the edges, about 10 minutes. Remove foil and weights and continue to bake until golden brown, about 10 minutes more.
- Brush entire inside of crust with egg yolk to seal and cool.

Filling:
2 lemons, zest removed and chopped
1 cup granulated sugar
2 large eggs
1/3 cup unsalted butter
1 9-inch tart shell, baked blind and brushed with egg yolk (the recipe above)

- Juice the lemons into a saucepan, add the zest, sugar, and eggs to the juice, and whisk to combine.
- Set pan over low heat and stir constantly to thicken slightly.
- Remove from heat and strain into a bowl.
- Stir in butter and pour into prepared tart shell.
- Bake until filling is jiggly in center, about 20 minutes.
- Cool on rack to room temperature, then chill and garnish as desired.

My friend, chef Cindy Goldman, comments that this is everything a delightful little lemon tart should be—sweet, tart, and pretty with a nice, crisp crust. Rosettes of sweetened whipped cream topped with candied lemon peel make a nice garnish. Precooking the lemon filling, as suggested in this recipe, will ensure that it sets properly.

If you look carefully, you may notice a change in the appearance of the eggs as you stir in the granulated sugar. Sugar dissolves in water, and since eggs are made partly of water, they incorporate the molecules of sugar. These sucrose molecules interact predominantly with the protein part of the egg: chicken egg white is a solution in water of a protein known as albumin (the Latin word for "white" is *album*).

Most people believe mistakenly that the sugar serves to compensate for the acidity of the lemon juice—for its tartness. What compensates for acidity is basicity, as in caustic soda, but you would ruin your innards introducing caustic soda into the preparation—and even with mere soda—that is, baking soda—your tart would be weird beyond description.

Some other interesting chemical reactions go on between the lemon juice and the sugar. The acidity from the lemon splits sucrose molecules into their components, glucose and fructose molecules. These molecules then become free to recombine. And they do so, in the heat of the oven. The resulting polymers of glucose and fructose monomers are known as caramel. As the baking proceeds, you will detect the onset of caramelization from the color of the surface; the longer you leave the tart in the oven, the darker it will get (since caramel has a brown color).

Similarly interesting chemistry involves the egg yolks and butter. Egg yolk consists of a lecithin, a chemical with predominantly fatty molecules: hence, it is an emulsifier of other fats, whether animal or vegetable (such as olive oil, which is how mayonnaise is prepared). Thus, the egg yolk-plus-butter part of the frothy mixture consists of a fatty emulsion.

The sweetened albumin part of the filling denatures in the heat of the oven, just like any good protein, which a temperature increase denatures irreversibly. This seals the pastry, not unlike the way a varnish seals a painting.

This is a pastry with an all-important yellow glow. A hint of caramelization suffices. The overall sunny look of the tart does not rule out decoration, such as small ribbons of lemon rind embedded in the filling. Soaking them first in Cointreau overnight, or candying them first, will make you appear quite a pastry chef, not to mention showing off your newly acquired knowledge of polymer chemistry!

A few other variants suggest themselves. You may want to try other fruits from the citrus family instead of lemons. You may even leave out the lemon juice, if you so desire, and replace the sugar in the above recipe with the same amount of honey: in honey, an enzyme contributed by the bees has already cleaved sucrose into a glucose-fructose mix, all set for caramelization under heat. You will then have reinvented *tarte au sucre*, an old-fashioned pastry still very popular with Belgians.

Speaking of national tastes—yes, there is such a thing—the French like both their tarts and their pizzas much drier than Americans do, with relatively more dough and less filling. Look at a fresh *baguette*: consider its optimal crust-to-ballast ratio. Do feel free to increase the amount of filling in the above recipe, by up to 50 percent. But don't blame me if you gain weight!

FOURTEEN

Make It Scarce?

Zumo de limón, zumo de perdición
(Lemon juice, juice of perdition)
SPANISH PROVERB

A dish, a dessert especially, combines a taste with a texture. Foams can serve the purpose admirably. The foodstuff, when inflated by tiny bubbles of air, acquires a light consistency. By dispersing volatile molecules, which partition themselves between the bubbles and the surrounding network, foams make the flavor more evanescent, thus all the more desirable on account of its newly acquired subtlety.

A delicious case in point is froth on top of a cup of espresso. The steam used in extracting the drink—and its aroma—from ground coffee beans caramelizes some of the sugar molecules from the coffee powder. Yes, coffee beans contain sugars, enough for such an effect. These caramelized sugars—polymers, as you will recall—form a network encasing tiny air bubbles, which in turn contain volatile aromatic molecules. As steam—water vapor—extracts the aroma from the powder and goes on to condense into hot liquid coffee, dissolved air rises to the surface in the form of the floating creamy froth.

Not only is it pleasing to the eye, it is also delicious, due to the dispersion of the molecules responsible for the coffee aroma—hundreds of different chemicals. When professional coffee tasters sample the quality of beans, they give much importance to the fine-celled foam that forms in a similar way upon pouring boiling water over freshly ground coffee. Cupping is the name of the tasting procedure.

The key phenomena here are dilution and phase separation. The latter owes its importance to inhaling. In order to be perceived, any fragrance must be volatile enough to get into the air we breathe. We do not smell the aromatic chemicals in the liquid coffee we drink, only in the atmosphere above it. Furthermore, the sense of smell is oversensitive. Our nose receptors get saturated in the presence of too many molecules. That is why we no longer smell an odor after having been exposed to it for a while. Paradoxically, any fragrance is all the more potent when it is diffused in smaller amounts. That is why perfumes are best applied as a mist, from a spraying device dividing the liquid into minute droplets. Other familiar examples include the "head" on a mug of beer and the mousse that forms when champagne or any other sparkling wine has been poured into a glass. What is true of the coffee aroma holds for other aromas too, such as those from citrus flowers or fruits.

———

To pursue the light-headed theme of foams, here is a recipe:

CITRUS SABAYON

3 pummelos
1 pint blueberries
1 pint raspberries
2 pints strawberries
8 egg yolks
3 tablespoons granulated sugar
1 cup champagne

- Carefully remove the citrus skin. Cut it into thin strips, leaving only the rind and discarding the white pith.
- Blanch about a fourth of a cup of rind strips in simmering water for three minutes.
- Excise the pith from the pummelo segments and remove the membranes. Arrange the segments and the berries on dessert plates.
- Mix together egg yolks, sugar, and champagne.
- Whisk the mixture together over a double boiler set over, but not touching, simmering water until the consistency thickens and ribbons form, about six minutes.
- Add the blanched rind strips (optional) to the mixture and spoon evenly over the fruit. Dust with powdered sugar and serve. Serves six.

As a rule, a sabayon is adorned with the flavor of an alcoholic drink—champagne in this case—which translates into numerous variants for this extravagant dessert. Extravagant? It combines a sweet with an after-dinner drink; served at the end of a good meal, its lightness contrasts with earlier and heavier courses. As far as desserts go, it is a virtuoso number on the part of the pastry chef.

But what is a sabayon? What are the reasons for the scrupulous temperature control? They are physicochemical in their simplicity. Do not shudder, please.

Egg yolk is mixed with sugar and beaten with a whisk, whose function it is to inflate the mixture with a host of minuscule air bubbles. Left on its own, such a preparation, of course, would deflate with time. It is held up by a network made of chains of natural detergent molecules and egg yolk lecithin, interspersed with sugar molecules, which forms the walls enclosing the tiny air pockets.

Its durability stems from the way it is heated; that is, from the partial cooking it undergoes. The role of the increased temperature is to speed up the assembly of the network, expand the air bubbles, and—no less important—to seal the envelope around each bubble. It is neither to evaporate the added alcohol—the sabayon is not inflated by alcohol vapor, only by air—nor to vaporize water—which constitutes fully 50 percent of the wine and the egg yolk—into steam. The sabayon is, truly, a gradually expanding hot-air balloon, not a steam engine. Accordingly, the temperature of the preparation ought not to go above about 154°F. What you want to avoid is denaturation of the protein in the egg yolk, resulting in cooked bits.

The name of this delicacy is nearly identical in the languages of its mother country, Italy, and of the seat of official cuisine, France. *Zabaglione* in Italian and *sabayon* in French sound very much alike. It was allegedly invented at the court of the Medici, in Florence, during the sixteenth century. I say "allegedly" because the etymology of the word relates to the name of a beverage, a kind of a beer, in Illyrian: *sabaia*. This would tend to place the origin on the Dalmatian coast, if not farther east. One may only conjecture that the Venetians imported the recipe from Asia, perhaps from one of their properties across the Adriatic. In any case, the etymology fingers the sabayon originally as a drink, like its cousin, eggnog, and probably one with the same purpose—hangover avoidance. Alcohol emulsified in the sugar and egg yolk combo is slower to enter the bloodstream across the stomach walls.

A sabayon owes its featherlike touch to the light, airy texture of a froth perfumed with a wine aroma. Marsala is a frequent partner in the symbiosis. Relatively recently, the food industry has made our palates familiar with such weightless sweets. In France, for instance, supermarket shelves bear substitutes for homemade chocolate mousses, large numbers of dairy products of the Fontainebleau type, chestnut puree whipped up into a mousse, and what have you. Prior to 1973 and the Cuisinart era, the sabayon was unique in its category of desserts.

Not that egg-based fluffy dishes and desserts were absent from menus. On the contrary, the training of a chef emphasizes technical prowess. Think of the various related sauces: hollandaise, choron, béarnaise, or maltaise (that is, a hollandaise flavored with orange juice). Chefs have to learn how to beat a raw material, such as heavy cream, egg white, or egg yolk, to the desired consistency. They are forcefully warned not to go beyond the desired emulsified stage, and they are indelibly imprinted with this technique by repeating it until it becomes instinctive. They learn how to beat egg whites into either soft peaks or, farther down the road, hard peaks. They know when to stop beating cream before it turns to butter. They are very familiar with the beating of egg yolks until the ribbon stage, which is the goal in making sabayon. I can think of at least one well-known restaurant, La Mère Poulard at Mont Saint-Michel, where the whisking of eggs has been a fixture since 1888, when it was introduced there. By now, diners watch the preparation as if it were a circus act or a performance on a late-night show. As they are led to their table, they pass a young chef's assistant, seated on a straw chair and holding a metallic bowl on his lap whisking eggs. This scullion is display-ing the hard work involved in preparing the eggs for the omelette that has made the establishment famous, unfortunately turning it into a tourist trap. The ensuing clatter, while not quite up there with the noise made by a rock band, is nevertheless memorable.

When preparing a sabayon, an artful display on the plate is essential. It hints at the contrasts that the mouth is about to experience, shaped citrus quarters and berries as opposed to the shapeless froth, dense and juicy fruit pulp in counterpoint to the airy and light mousse, tart citrus and berries versus the sweetness of the sabayon.

Quite a few variants of this recipe exist. Pieces of pummelo are most often featured. However, many other citrus fruits can be tapped for the role. An elegant and attractive substitution is smaller fruits such as cala-mondins or kumquats. Calamondins are reminiscent of tangerines but much smaller, about an inch in diameter. Their name, from *kalamunding* in Tagalog, indicates that calamondin trees are native to the Philippines. They are not hard to find in the United States; they have been grown in Florida since the turn of the twentieth century. Often, a grower will include a branch of calamondins in a shipment of citrus for decoration. Many Asian food stores carry them. Their peak season is November through June. Calamondins are best kept in the refrigerator and should be used within a week. There are five to nine segments around a small semi-hollow axis. They peel easily, and the flesh is orange, juicy, and acidic.

Kumquats are another choice for substitution. Their name is an Anglicization of "Kin ku," meaning "golden orange" in Chinese. Robert Fortune, who collected plants in China for the London Horticultural Society, brought back kumquat cuttings and gave them their English name. Kumquats are related to calamondins. Some botanists hold the latter to be hybrids of kumquats and mandarins. Kumquats are best bought when bright orange. Just as with calamondins, the peel should be of uniform color; the fruit is best purchased when firm, with a lemonlike firmness. Like calamondins, kumquats peel easily, and they also have a thin skin. For that reason, they are more perishable than a lemon, a grapefruit, or a navel orange.

––––––––

Another example of a frothy delight is the orange mousse. *Musse de laranja* is the Portuguese name of this dessert, which is shared by traditional cuisines in both Portugal and Brazil.

ORANGE MOUSSE

3 eggs
1 cup granulated sugar
The grated peel from 1 orange (carefully washed first, to remove all traces of
    pesticides)
1/2 cup orange juice
The grated peel of 1 lemon (also carefully washed) and, separately, the juice
    therefrom
1 cup heavy cream
2 cups (0.25 L) milk

• Separate the yolks from the whites.

In Portuguese, while the latter are named *claras*, "the clear ones," for their transparency, the former are known as */gemas*, "the gems," for their color.

• Beat the yolks together with half the sugar for 10 minutes.

193

The whipping produces a foam, a *mousse* in French: that is to say, an ensemble of tiny air bubbles in a network of molecules. Beating either egg yolks or egg whites with a whisk lets in air, and thus turns them white. This effect is caused by white light scattered by all the microscopic air bubbles in the preparation.

Sweetening the eggs is necessary to produce both a sweet taste and a significant caloric intake. This part of the recipe reinforces the sugars already present in the orange and lemon juices. It will boost the sugar-to-acid ratio in the juices.

- Boil the milk in a saucepan for 10–15 minutes until it is reduced by half, then let it cool. Slowly add the beaten and sweetened egg yolks. Heat the preparation gently, while continuing to beat it steadily to help maintain the foam. Add a pinch of flour, while continuing to whisk. As soon as the mixture has thickened, remove the pan from the heat and let it cool.

This stage serves to stabilize the mousse in two ways. Heating generates microbubbles of water vapor, which get trapped in the existing network made up of the molecules of fat (lipids) and sugar (carbohydrate) in the beaten egg yolks, which are cooked ("burnt," as cooks say) by the earlier incorporation of sugar. The giant starch molecules from the flour prevent aggregation of this custard into globs. The macromolecules of starch strengthen the preexisting network, trapping the bubbles of gas.

- Though it is optional, further thickening of the mousse can be done with three or four leaves of gelatin, first soaked in cold water.
- Mix in the citrus juices and the grated orange and lemon peel. Refrigerate for 20 minutes.
- Meanwhile, beat the egg whites with a whisk or an electric beater to firmness. Incorporate the rest of the sugar and proceed with the whipping until you have achieved a hard and shiny consistency.

The purpose of beating is, as in the foam prepared from the yolks, to turn the egg whites into a stable foam, with the air bubbles enveloped by the protein molecules, known as ovalbumin, of the egg whites.

- Whip the heavy cream. Mix in the cooled pale yellow mixture and the egg whites. Set your ensuing orange mousse in the refrigerator for several hours before serving.

Grapefruit juice can replace orange juice in a variant of orange mousse. The word "mousse" is mouthwatering in itself. It comes by way of the French language from the Latin name for honey, *mel*. One of the derivatives of *mel* is *mulsum*, the name for a wine admixed with honey. Taken as a term of affection, as Americans call a person they like "honey," Romans in antiquity would name their tender one *mulsa*.

———

An orange mousse or a citrus sabayon is an excellent dessert to enjoy at the end of a *feijoada* dinner. The *feijoada* (pronounced "fay-jo-ada"), named for the beans (*feijão*) in it, has become the national dish of Brazil. It is comparable, perhaps, to such dishes as a French bouillabaisse, a German sauerkraut, a Scottish haggis, or an English leg of lamb with mint sauce. Brazilians have turned a quotidian dish, prepared on the plantations by African slaves, into a meal emblematic of their whole culture.

FEIJOADA

4 lbs dried black beans (Goya brand, for instance)

2 lbs carne seca (Brazilian salted cured beef)

2 lbs salted pork meat (pork butt or shoulder, trimmed of fat, will also do)

1 lb hot sausage (Spanish or Portuguese chorizo will do)

1 lb spareribs

1 lb salted pork loin

1 pig foot

1 pig ear

1 pig tail

1 pig tongue

1/4 lb unsmoked pork fat (bacon)

3 cloves garlic, crushed

4 tablespoons minced onion

1 bay leaf

3 tablespoons olive oil

- Wash the beans. Soak them overnight in cold water.
- Wash the meats well to remove the excess salt. Soak them also in cold water, but in a separate bowl, changing the water from time to time.
- The following morning, place the beans and the meats in separate saucepans, with water to cover by 3 inches, and cook over medium heat. You will need a large saucepan for the beans. Do not add salt to the beans; they would harden irreversibly. After a couple of hours of cooking, the beans will have softened. You can also use a pressure cooker for the beans, in which case their cooking takes only 40 minutes.

- In a separate saucepan, heat the olive oil, add the onion and the garlic, and cook until golden brown. Add three or four tablespoons of beans, mash them and mix well. Add a few tablespoons of the bean juice and let this mixture come to a boil. Then, put it in the pan together with the beans. It will thicken and season them.
- When the meats are cooked, throw them into the saucepan with the beans. Let this mixture simmer for another 15–20 minutes, so that all the ingredients mingle their flavors. If you are fond of laurel, or rather of its taste, put in a bay leaf. Add salt, and if necessary, add more. But do not forget that the meat is already salted!
- For serving, use earthenware bowls and/or plates. On individual plates, place both beans and the meats cut into small chunks. The feijoada is served together with white rice, cabbage (see below), toasted manioc meal known as farofa, a half-dozen medium-acidic oranges, peeled and cut into thin slices, and its own sauce. (Serves ten to twelve.)

SAUCE

1 cup lemon juice
3 tablespoons finely chopped parsley and chives
1 tablespoon chopped green pepper
1 pinch chili powder
Salt

- Mix the peppers and the other ingredients according to taste.

CABBAGE (COUVE À MINEIRA)

3 heads of cabbage
2 tablespoons chopped onion
4 tablespoons olive oil
1 teaspoon salt

- Wash the cabbage, remove the stems, set the leaves on top of one another, roll into a bundle and slice them finely. Remove the bitterness by pressing in a bowl with cold water.
- Brown the onion in the oil. Drain the cabbage and add it. Season with salt and mix well. Leave a few minutes on medium heat, mixing well. It should be cooked al dente.

There are as many recipes for *feijoada* as there are Brazilian cooks! There are also regional variants. The recipe above is from Rio de Janeiro. Both the *carioca feijoada* and the *gaucha feijoada*, from southern Brazil (the state of Rio Grande do Sul), use black beans. In the interior state of Minas Gerais, the *feijoada* is made with brown beans instead—which makes for significantly faster cooking. Do not let yourself be overwhelmed by the long list of ingredients. Some can be omitted, and you will still have a yummy dish.

A Brazilian family will eat a *feijoada* on a weekly basis, traditionally on a Saturday afternoon, when they have company and there is time to enjoy the meal. If someone brings a guitar to pluck and for the diners to sing to, all the better! You should select your own favorite drink to go with this dish. Brazilians will drink a *caïpirinha* or two, toward the beginning of the meal perhaps, followed by beer.

The *feijoada* has an African origin. It was, during colonial times, what the slaves on a plantation cooked for themselves and ate. Imagine a large cast iron pot simmering on a fire in the backyard, with the cook throwing in those pieces of meat she would have been able to obtain, typically the throwaways from the kitchen of the masters, especially cheap cuts of pork.

———

To me, the visual aspect of the *feijoada* serves as a reminder of the human mix of which Brazil is made. The white rice stands for the Portuguese colonists, the black beans for their African slaves, and the sliced oranges for the Indian component—which, alas, has mostly disappeared nowadays.

A good *feijoada* is a feast for the mouth because it presents it with a wide variety of textures and tastes: the powdery and somewhat astringent *farofa* contrasts with both the grainy rice and the fibrous meats; the creamy beans contrast with the juicy orange slices. Indeed, the orange slices are crucial to the enjoyment of this dish: their tartness serves as a counterpoint to the mealy flavor of the beans. The *feijoada* arouses the sensuality of the mouth as a whole, exciting all of the nerve endings with sweet and spicy flavors, lending itself to appreciation of tastes both raw and cooked.

———

To do justice to the diversity of citrus, I will close this chapter with a less familiar fruit. The Ugli (pronounced oogly) richly deserves its name. It is a nondescript fruit of roughly spherical shape: it looks like a used and crumpled paper bag. In color, it is likewise unsettling, green with patches

197

of dull orange, or vice versa. Even the spelling of its name is daunting, both because of the final "i" instead of the expected "y," and because of the registered trademark sign that often flanks it.

Anyway, what's an Ugli? It is a relatively recent discovery—even though it may have been part of nature for centuries. In (or about) 1917, G. G. R. Sharp, who owned a Jamaican estate, Trout Hall, was walking about in a pasture when he came upon a strange citrus tree, bearing an unseemly fruit that resembled—if one can imagine such a monster—both a tangerine and a grapefruit.

Sharp collected some of the budwood and grafted it onto an orange tree. He repeated the procedure, regrafting the progeny, using as his criterion for selection the number of seeds—opting for the fewest. By 1934, Sharp was exporting the fruit to England and Canada. He started supplying the U.S. market in 1942. The fruit is available, generally speaking, from October until February. At some point, Trout Hall Estate registered the name Ugli, which is thus uniquely associated with this single plantation.

The Ugli, grapefruit-like, is a large fruit, ranging from four to six inches in diameter. Tangerine-like, it is easy to peel. And, as befits a tangelo, it has a lovely taste, combining the sweetness of the tangerine with the tanginess of the grapefruit.

If you have always aimed at becoming more voluble, and, Ulysses-like, more nimble of tongue, here is a tongue twister (to be repeated loud, a half-dozen times in succession):

The monkey on the Eskimo's shoulders ogles the ugly Ugli in the igloo.

Did you succeed? Congratulations. As your richly deserved reward, you might consider feasting on the following:

UGLI DUCKLING

1 duck (about 4 lbs)
1 Ugli
1 teaspoon blade mace
1/4 pint (150 mL) dry white wine
1/2 teaspoon black peppercorns

- Prick the duck all over with a fork and rub a little salt into the skin.
- Place the duck in a roasting pan and bake at 375°F, without basting, for 1–11/2 hours, till cooked and nicely golden brown.

- Remove the segments from the Ugli. Cook them for no more than 3 minutes in the white wine, seasoned with the peppercorns and mace. Pour this sauce over the duck after carving and prior to serving. (Serves four.)

———

The introduction of some of the more exotic citrus fruits—calamondins, desert limes, kumquats, the Ugli—in these last two chapters is an appropriate note on which to end this book. Exoticism not only excites the imagination, but also leaves room for thought.

Citrus trees of all varieties were westward immigrants from Asia. They came first to the Middle East and then moved on to North Africa and western Europe. From the sixteenth century on, they were transplanted to the West Indies and the Americas. These plant species are vulnerable to frost; acclimatization did not make them hardier. And citriculture, when a monoculture, becomes a breeding ground for many a plant disease.

Are we in danger of losing our sense of wonder? When I was a child, oranges were wonderful. They and their kindred, mandarins and tangerines and clementines, owed their prestige to their scarcity. I would see them and eat them only at Christmastime. Now, in our age of mass production and mass distribution, when the planet has turned into a global supermarket, when tourism breeds uniformity, when multinational corporations (Coca-Cola, McDonald's) add addictive unhealthy diets, citrus fruits in their diversity threaten to become mundane. We have even become used to sampling in restaurants tiny citrus fruits we have rarely seen before. Recipes such as the ones in the preceding pages, calling for kumquats or calamondins or key limes, are becoming commonplace.

Spoiled? Assuredly. When we hear about seasonal foods from Laos, such as crickets, green tree ant eggs, river algae, and wild cassava, becoming choice items for North American chefs, or morsels such as cassava crisps turning into snacks for Californians, a sense of transgression and outrage may be the result. Not so much out of shame, in that we are taking food for our jaded palates and our overfed stomachs away from needy mouths, but because, more selfishly yet, we are losing the Garden of Eden by partaking too much of it. Is it not morally wrong to deprive ourselves of delight in certain foods by turning them into ordinary household items?

In Western culture, for many a century, citrus was a luxury. How can we contrive to keep it such?

Mother Nature may help us. There are thousands of existing varieties of citrus, and genetic engineering now helps us to devise many more, and thus to broaden the range of trees from which to partake. We are at the stage when we can design our own Garden of Eden. Many exotic and wonderful fruits are within reach.

Citrus is all the more attractive for having such a pliant genome. There may be no end of delights in sight, and it is to be hoped that the aura of luxury surrounding citrus fruits will not be lost.

Epilogue: Answer from a Fellow Writer

Honored Colleague:

To receive your letter was, frankly speaking, a surprise. I had not realized that my book on citrus trees and fruit, *Chü Lu* (The Orange Record), would find any readership, let alone that it would outlive me. Had I suspected such a destiny, I would have drafted it yet more carefully. My penmanship does not, to my taste, always express my thoughts accurately enough. I feel distressed that the administrative style of my reports may have crept into the text of my book.

Thank you also for submitting to me your own book. I found it interesting at times, and often surprising. May I bring to your considered attention the cursory treatment of and short shrift you often give to fruits important in our part of the world, such as, to mention just two, calamondins and kumquats? I realize that you are writing for a different readership. However, both you and I know, as writers, that a book, any book, creates its own resonance and following. I do not understand why you would even consider catering to whims and prejudices, which will soon fade away like mist in the morning sun. If I may venture, with considerable misgivings, a personal question, why did you not indulge in self-expression a great deal more? Could it be because you govern a province as I used to do, and you have to submit to administrative restraints as I have had to? I have no idea what your occupation is.

My lack of surprise may well surprise you. That citrus plants have spread far and wide is something to be expected.

Our Empire of the Middle knows well how to export its inventions, its cultural innovations—in short, its civilizing influence. During my lifetime I witnessed, to mention just one example, the enthusiastic adoption of our celadon stoneware by the Koreans, who made it their own. In this connection, a part of your book that was fascinating to me was your description of the extension of the Silk Road beyond the western ocean, into the Land of the Glowing Embers.*

As your senior, also now as a ghost in the sphere of the spirits, I view my responsibility as not to cast a shadow. I do not wish to influence the living. I have laid down my brushes, and my block of ink has presumably returned to the good earth.

In my present phantasmal state, the little that remains of my insignificant will hopes only to add to the translucence and limpidity of all things alive—including writings, spoken words, and reverence for one's ancestors. If I may paraphrase our great Tao philosopher Chuang Tzu, and I understand that some among Western philosophers in your own time have uttered similar paradoxes concerning the meaning of life:**

Am I Han Yen-Chih, fancying myself an orange blossom,
Or am I an orange blossom fancying itself as Han Yen-Chih?

I wish your book, Honored Colleague, similar lightness of spirit.
Yours truly,
Han Yen-Chih

*Brazil
**Perhaps he is looking forward to Alfred North Whitehead.

Selected Notes

Unless otherwise noted, all translations in these notes are my own. A full set of notes can be accessed on my Web site, http://www.pierrelaszlo.net. Readers will find here a bibliographic essay whose main features are a readable set of notes, sometimes in a conversational mode, a list of suggestions for further reading, an atlas of exploratory moves in directions this book could not cover, and explanations for some of the author's interests.

PROLOGUE

Chü Lu (The Orange Record): H. Yen-Chih, *Chü Lu* (*Monograph on the Oranges of Wên-chou, Chekiang*) (Leiden: E. J. Brill, 1923).

Mencius (ca. 372–289 BC): S. Couvreur, ed., *Les Quatre Livres. Oeuvres de Meng Tzeu* (Leiden & Paris: E. J. Brill and Les Belles Lettres, n.d.), 456 (my translation.)

the hobby of raising miniature plants: The bonsai craft was indeed practiced in China at the time of the writing of the Orange Record, in the twelfth century.

CHAPTER 1

characteristics of the genus *Citrus*: M. Gaskins, "Understanding Citrus Fruit Growing," Technical Paper 16 (Arlington, VA: Volunteers in Technical Assistance, 1984).

the interrelationshops between citrus types and varieties: There are four main types of sweet oranges (*Citrus sinensis*): common, sugar or acidless, navel, and blood orange. There are at least four types of mandarins: common mandarins (*Citrus reticulata* Blanco), King mandarins (*Citrus nobilis* Loureiro), Mediterranean mandarins (*Citrus deliciosa* Tenore), and Satsuma mandarins (*Citrus unshiu* Marcovitch). The various tangerines are *Citrus reticulata* varieties.

also known as the Feast of the Tabernacles: see "Annie's Feast of Tabernacles Page," http://www.annieshomepage.com/tabernacles.html, for comprehensive information. The midrashic interpretation of the symbolic analogy of the citron to the human heart is mentioned in Asoph Goor and Max Nurock, *The Fruits of the Holy Land* (Jerusalem: Israel Universities Press, 1968), 155.

Islam established a major beachhead: L. P. Harvey, *Islamic Spain, 1250 to 1500* (Chicago, IL: University of Chicago Press, 1990); Richard Fletcher, *Moorish Spain* (London: Weidenfeld & Nicolson, 1992); Hugh Kennedy, *Muslim Spain and Portugal: A Political History of al-Andalus* (New York: Longman, 1996); Olivia Remie Constable, *Trade and Traders in Muslim Spain: The Commercial Realignment of the Iberian Peninsula, 900–1500* (Cambridge: Cambridge University Press, 1994); T. F. Glick, *Irrigation and Society in Medieval Valencia* (Cambridge, MA: Harvard University Press, 1970).

an aristocratic house in Braga: Maria da Asunção Jacome de Vasconcelos, "A Casa das Coimbras," *FORUM* 18 (1995): 63–80.

a nineteenth-century description of the convent: Padre António Carvalho da Costa, *Corografia Portugueza e Descripção Topografica* (Braga: Domingos Gonçalves Gouvea, 1868).

The bergamot orange is a case in point: For the trajectory of the bergamot orange, see A. di Giacomo and C. Mangiola, *Il Bergamotto di Reggio Calabria* (Reggio Calabria, Italy: Laruffa Editore, 1989); R. Huet, A. B. N'Zie, and R. Dalnic, "Etude comparative de l'huile essentielle de bergamote provenant d'Italie, de Corse et de Côte d'Ivoire," *Fruits* 36, no. 6 (1981).

another orange variety: For more on blood oranges, see D. Karp, "The Blood Orange," *Saveur*, January 1997.

Jacques Prévert: from *Œuvres Completes*, 2 vols. (Paris: Gallimard, 1992, 1996) (my translation).

Clementines are named for Father Clément Rodier: L. Trabut, "Une nouvelle tangerine, la clémentine," *Bulletin de la Direction de l'Agriculture* (*Gouvernement Général de l'Algérie*) 35 (1902): 21–35; L. Trabut, "L'hybridation des citrus: Une nouvelle tangerine, la clémentine," *Revue d' Horticulture* (Paris) 74 (1902): 232–34; L. Trabut, "La clémentine, les hybrides du *Citrus nobilis*," *Direction de l'Agriculture et de la Botanique* (Algérie) 67 (1926).

made from Malbec grapes: M. Lonsford, "Mendoza, Malbec, 'Magnifico': Region's Little-Loved Grape Becomes a Popular Pick," *Houston Chronicle*, August 20, 2002.

As Fernand Braudel pointed out: Fernand Braudel, "European Expansion and Capitalism: 1450–1650," in *Chapters in Western Civilization* (New York: Columbia University Press, 1961), 245–88.

the Canaan Sage: N. Wachtel, *La foi du souvenir: Labyrinthes marranes* (Paris: Le Seuil, 2001). See also M. Kriegel, "Le marranisme: Histoire intelligible et mémoire vivante," *Annales HSS*, mars-avril 2002, 323–34. The largely lost figure of *O Bacharel de Cananéia* is tempting to the novelist; indeed, several semi-fictional books exist, such as J. R. Torero and M. A. Pimenta, *Terra Papagalli* (Rio de Janeiro: Editora Objetiva, 1997) and José Pereira da Graça Aranha, *Canaã* (Rio de Janeiro: Ediouro, 2002 [1902]).

Rodolfo Garcia: C. S. Moreira and S. Moreira, "História da citricultura no Brasil," in *Citricultura brasileira*, vol. 1, ed. O. Rodrigues and F. Viêgas (Campinas, SP: Fundação Cargil, 1980), 1–21. Manuel da Nóbrega, a Jesuit, in a letter he wrote in 1549, described citriculture in Bahia: Edmundo Navarro de Andrade, *Manual de citricultura, cultura e estatistica* (São Paulo: Chacáras e Quintais, 1933).

Jean de Léry: The testimony of this traveler, one of the founding figures of ethnography, comes from his published diary: J.-C. Morisot, ed., *Jean de Léry, Histoire d'un voyage fait en la terre du Brésil* (Genève: Librairie Droz, 1975 [1578]).

Brazilian slavery differed not in kind, but in degree: See E. D. Genovese, *The World the Slaveholders Made* (New York: Pantheon Books, 1969); L. Foner and E. D. Genovese, eds., *Slavery in the New World: A Reader in Comparative History* (New York: Prentice Hall, 1970).

Jean Estiemble: *Généalogie et Histoire de la Caraïbe* 27 (May 1991), 327.

a certain Captain Shaddock: This is not a fictional figure. J. Kumamoto, "Mystery of the Forbidden Fruit: Historical Epilogue on the Origin of the Grapefruit, '*Citrus paradisi*' (Rutaceae)," *Economic Botany* 41 (1987): 97–107.

Its first recorded mention: R. W. Scora, J. Kumamoto, R. K. Soost, and E. M Nauer, "Contribution to the Origin of the Grapefruit, *Citrus paradisi* (Rutaceae)," *Systemic Botany* 7, no. 2 (1982): 170–77.

The grapefruit was introduced into Florida: For the trajectory of the grapefruit, see R. W. Hodgson, "Horticultural Varieties of Citrus," in *The Citrus Industry*, ed. W. Reuther, H. J. Webber, and L. D. Batchelor (Berkeley: University of California, Division of Agricultural Sciences, 1967), 431–591; Julia F. Morton, "Grapefruit," in *Fruits of Warm Climates* (Miami, FL: J. F. Morton, 1987), 152–58.

Sour limes originated: For the trajectory of the sour lime, see Hodgson, "Horticultural Varieties of Citrus"; Morton, "Mexican Lime," in *Fruits of Warm Climates*, 168–72.

Joseph Sturge: Stephen Hobhouse, *Joseph Sturge: His Life and Work* (London: J. M. Dent and Sons, 1919); Alex Tyrrell, *Joseph Sturge and the Moral Radical Party in Early Victorian Britain* (London: Christopher Helm, 1987).

the key lime: See R. L. Phillips, S. Goldweber, and C. W. Campbell, "Key Lime," Fact Sheet FC-47, Fruit Crops Department, Florida Cooperative Extension

Service, Institute of Food and Agricultural Sciences, University of Florida (February 1991; revised April 1994). I can also recommend the attractive Web site http://keylime.com.

Valencia oranges: Hodgson, "Horticultural Varieties of Citrus"; J. Saunt, *Citrus Varieties of the World* (Norwich, England: Sinclair International Ltd., 2000); W. P. Bitters, L. A. Batchelor, and F. J. Foote, "Valencia Orange Strains," *California Citrograph* 41 (1956): 277–79, 283–84; B. O. Clark, "Introduction of the Valencia Orange into California," *California State Fruit Growers Convention Proceedings* 48 (1916): 39–43.

the Jaffa sweet orange: Hodgson, "Horticultural Varieties of Citrus."

The navel variety of orange: Do royal weddings contribute to the advancement of science? In 1817, the Archduchess Leopoldina traveled from her native Austria to Brazil to wed Dom Pedro de Alcantara, the future emperor of Brazil. She was accompanied by an Austrian commission, which included two naturalists, Johann Baptist von Spix (1781–1826) and Karl Friedrich Philip von Martius (1794–1868). Their expedition within Brazil lasted nearly three years. They returned to Europe in June 1820, having traversed central Brazil into the Amazon region. Their collection was most impressive: they brought back to Munich 85 species of mammals, 350 birds, 130 reptiles and amphibians, 116 fishes, 1,800 coleopters, 120 orthopters, 30 neuropters, 120 hymenopters, 120 lepidopters, 250 hemipters, 100 dipters, 80 arachnids, several crustaceans—and a description of the *umbigo* orange. Jean-Baptiste De-bret (1768–1848), a student of David who trained at the Ecole des Beaux-Arts in Paris, went to Brazil as a member of the French Artistic Mission for two royal events. He contributed to decorating Rio de Janeiro in 1816 in antici-pation of the coronation of Dom João VI, king of Portugal. He was still there for the arrival of archduchess Leopoldina. Establishing himself in Brazil, he made numerous documentary watercolors and lithographs. He returned to Paris in 1831, after having been elected to the Académie des Beaux-Arts. The book he published is invaluable for its visual documentation of Brazil during the first third of the nineteenth century: J.-B. Debret, *Voyage pittoresque et historique au Brésil, ou Séjour d'un artiste français au Brésil depuis 1816 jusqu'en 1831 inclusivement* (Paris: Firmin-Didot frères, 1834–1839). An official USDA publication still gave the name of the fruit in 1917 as the "Bahia navel": P. H. Dorsett, A. D. Shamel, and W. Popenoe, "The Navel Orange of Bahia; with Notes on Some Little-Known Brazilian Fruits," *U.S. Department of Agriculture Bulletin* 445 (1917): 35.

Two Germans: J.-B. von Spix and C. F. P. von Martius, *Reise in Brasilien auf Befehl Sr. Majestät Maximilian Joseph I. Königs von Baiern in den Jahren 1817 bis 1820 gemacht von Weiland*, 3 vols (Munich: M. Lindauer, 1823–1831).

Luther and Eliza Tibbets: M. T. Mills, "Luther Calvin Tibbets, Founder of the Navel Orange Industry," *Historical Society of South California Quarterly* 25 (1943): 126–61; J. Hunneman, "A New OJ King," *North County Times*, January 18, 2002.

William Saunders: A native of St. Andrews, Scotland, he was raised in a family
of gardeners. He studied horticulture in St. Andrews and at the University
of Edinburgh. In 1848, he brought his bride to the United States. They first
settled in New Haven, Connecticut. After forming a partnership in 1854
with Thomas Meehan in Philadelphia, he designed the Clifton Park estate
of Johns Hopkins and the Ross Winans place in Baltimore. With the first issue of *The Farmer and Gardener*, a prize was offered for the best essay on the
grape, which Saunders won: W. Saunders, *An Essay on the Culture of the Native
and Exotic Grape* (New York: J. B. Lippincott & Co., 1860). He was appointed
Superintendent of Horticulture at the U.S. Department of Agriculture in
1862. In 1863, he designed the National Cemetery at Gettysburg. Saunders
believed that "the prevailing expression . . . should be that of *simple grandeur*.
Simplicity is that element of beauty in a scene that leads gradually from one
object to another, in easy harmony, avoiding abrupt contrasts and unexpected features. Grandeur . . . is closely related to solemnity." Two years later,
General Grant picked Saunders to select the site for and design the Lincoln
Monument at Springfield, IL. In Chicago, he designed the Rosehill Cemetery. For his philosophy of landscape gardening, see *Landscape Gardening*, Report of the Commissioner of Agriculture for the Year 1869 (U.S. Department
of Agriculture, 1870). See also William Saunders, *Report of the Superintendent
of Gardens and Grounds*, Report to the Commissioner of Agriculture for the
year 1892 (U.S. Department of Agriculture, 1893).
Frank N. Meyer: I. S. Cunningham and F. N. Meyer, *Frank N. Meyer: Plant Hunter in
Asia* (Ames, IA: Iowa State University Press, 1984).
"I am pessimistic by nature": All quotations from Frank Meyer are from the Plant
Explorers Web site: http://www.plantexplorers.com/explorers/biographies/
meyer/franknicholas-meyer.htm.
David Fairchild: D. Fairchild, *The World Was My Garden: Travels of a Plant Explorer*
(Miami, FL: Banyan Books, 1982).
Walter Tennyson Swingle: H. B. Humphrey, "Makers of North American Botany,"
Chronica Botanica 21 (1961): 10–15; E. B. d. Varona, "Walter Tennyson
Swingle (1871–1952) Collection Register," catalog of holdings (Archives
and Special Collections, University of Miami Libraries, n.d.), 5.
A Texas grower: R. W. Apple, Jr., "Ten-Gallon Grapefruit: Living Up to Texas Leg-
end," *New York Times*, March 7, 2001.

CHAPTER 4

so-called limonaie: Note the discrepancy between the Italian term, giving primacy to protection of lemon trees, and the French (*orangerie*) and the English
usages (orangery), which stress orange trees instead.
many a villa built in Tuscany: such as Villa San Michele, in Florence, today turned
into a hotel. The *Limonaia* dates back to 1752. It now houses the best suites.
Villa Sasseto is now a summer rental. Villa Palmieri is located on the old road

between Florence and Fiesole. Boccaccio's *Decameron* (ca. 1350) mentions its idyllic garden with flourishing "bright green orange and lemon trees," M. Woods and A. S. Warren, *Glass Houses. A History of Greenhouses, Orangeries and Conservatories* (New York: Rizzoli, 1988), 5.

Charles VIII: In 1495, Charles VIII brought back to France in his escort Pasello de Mercogliano, from Naples, who built him an orange garden for the castle at Amboise, by the Loire River. Mercogliano built other, similar gardens for King Louis XII at his château in Blois and for Cardinal Amboise in Gaillon, as well as many others. King François I had orange trees in tubs surrounding a fountain to Diana in front of his castle at Fontainebleau in the 1520s. Woods and Warren, *Glass Houses*, 8.

A Medici from Florence: Some authors claim erroneously that the Medici crest, depicting a number of red balls on a gold shield, shows oranges. These balls, *palle* in Italian, actually refer either to pharmaceutical pills or to coins, in a by-now-forgotten allusion to the beginnings of this powerful family as either pharmacists or money changers. Rudolph Modley, Diana G. Comer, and William R. Myers, *Handbook of Pictorial Symbols* (New York: Dover, 1976).

Maria da Medici: She had the garden architect Francini design renovated gardens for the castle at St. Germain-en-Laye, west of Paris. She ordered orangeries for the gardens at the Tuileries, at Palais du Luxembourg, and at Fontainebleau. Woods and Warren, *Glass Houses*, 19.

the gardens of Florence: Those at Villa Aldobrandini are the obvious model for the St. Germain-en-Laye castle. A *limonaia* (Citrus Hall) existed at the Boboli Gardens built by Zanobi del Rosso in 1777–1778. See http://www.ilmisterodellagenesi.com/Limonaia.html.

a genius of advertising: *Louis XIV, Manière de montrer les jardins de Versailles* (Paris: Plon, 1951).

bringing water to Versailles: By 1668, the daily water requirements at Versailles exceeded those of the entire population of Paris, then 600,000 strong. To bring water from the Eure River, a gigantic project was planned at Maintenon (1684). It called for forty miles of canals and aqueducts. Construction was abandoned after ten years. The Marly machine, built in 1688 and still operating, consists of fourteen huge water wheels, lifting water five hundred feet above the Seine River, and has a daily output of 5,000 cubic meters of water. Georgia Santangelo, ed., *Les maîtres de l'eau, d'Archimède à la machine de Marly* (Paris: Musée promenade de Marly/Louveciennes/Art Lys, 2006).

an agricultural deity: Mars was originally invoked as guardian deity of cultivated fields and pastures. The Indo-European "clear-sky" gods with whom Mars must be equated were all, it seems, originally protector gods. In early Roman history he was a god of spring growth and fertility and the protector of cattle. Ovid also called him the "god of husbandry, of shepherds and seers." The name "Mars" has been related to Indo-European words for "boundary," "border," and so on: Gothic *marka*, Old English *maere*, Latin *margo*, Sanskrit *maryada*, Avestan *maraza*, Hittite *mark*, and so forth. The Latin *margo* means "edge, border, margin, boundary, shore," and is the cognate of English words such as "margin" and "demarcation." Thus Mars could have been so named as the guardian of field boundaries. In the old Roman calendar, March was the first month of the year, and thus demarcated the calendar. A. Cauquelin, *L'invention du paysage* (Paris: Presses universitaires de France, 2000), 117–18; G. Dumézil, *Jupiter, Mars, Quirinus: Essai sur la conception indo-européenne de la société et sur les origines de Rome* (Paris: Gallimard, 1941).

a long nave: twice the length of the original Le Vau building (1664), 13 meters high, with a 2-meter-deep roof. P. Bourget and G. Cattani, *Jules Hardouin-Mansart*, Les grands architectes (Paris: Vincent, Fréal et Cie, 1960).

The domesticated trees: Louis forced an extensive vertical growth of the naked trunk, crowned with a spherical growth of fruit-adorned branches. A compelling analogy for these giraffe-like trees with a long neck is the hyperbole, one of the classic tropes in rhetoric. P. Fontanier, *Les Figures du discours* (Paris: Flammarion, 1993 [1827]).

when he assumed real power: *La Prise du Pouvoir par Louis XIV*, directed by Roberto Rossellini (Paris, 1966).

Marc Bloch: in his *Seigneurie française et manoir anglais*, Cahiers des Annales 16 (Paris: Armand Colin, 1960).

English manors: John Evelyn, the diarist and polymath, also an architect (and a close friend of Christopher Wren), after traveling extensively on the Continent, published an ingenious scheme for heating the air outside a conservatory. The Grange, which Charles Robert Cockerell designed in 1824 and built in Hampshire in 1825, was endowed with a roof designed to reduce heat loss. The iron barrel vaulting had two layers, two inches apart. An influential building, it was featured in the *Gardeners Magazine* in both 1826 and 1827, and Charles Macintosh praised it in *The Greenhouse, Hot-House and Stove* (London: W. S. Orr, 1838). It owed its influence to its transitional character, melding the traditional brick and stone and the newer wide expanses of glass. "Cockerell, Charles Robert 1788–1863," in *Chambers Biographical Dictionary* (Edinburgh: Chambers Harraps Publishers Ltd, 1997).

turned into barns or stables: The orangery at Hotel Le Peletier de Saint-Fargeau, in Paris, was built in 1690. During the nineteenth century it was used for workshops. P. Velay, "Ouverture prochaine de l'orangerie," *La Revue du Louvre et des Musées de France* 50, no. 4 (2000): 25.

The one at Schönbrunn Palace: Austrians claim it as the second largest baroque orangery in the world, after Versailles. The first orangery, with a hothouse to winter bitter orange trees, dates back to the time of the dowager empress Wilhelmine Amalie. In 1754, Franz I. Stephan had Nicola Pacassi build (probably on designs by Nicolas Jadot) the building that still stands, 189 meters long and 10 meters wide. In 1786, Mozart conducted his Singspiel "The Impresario" there. It is now used for the Schönbrunn Palace concert series; see http://www.imagevienna.com/english.

Claude Monet's *Nymphéas*: On November 12, 1918, Claude Monet wrote to Georges Clémenceau, offering to the French state "two decorative panels." "It's not much," he added, "but it's the only way I have of taking part in the Victory." The two panels eventually (1927) became the twenty-two that make up the "Water Lily" murals installed in the Orangerie. J. Fayard, "The Gallery: A Veritable Ocean of Monet's 'Water Lilies,'" *Wall Street Journal*, July 1, 1999, A21; A. Riding, "Monet's Endless Valedictory of Light and Waterlilies," *New York Times*, June 1, 1999, 2.

from capital cities in Asia: Will archeological finds bolster this hypothesis? There is some supporting evidence. Nara was the capital of Japan, in the eighth century, on a Chinese model and plan, that of Chang'an, the capital of Tang China. Willows, mandarin orange trees, and locust trees were planted as roadside trees, in strict alignment. Such trees are mentioned in the *Man'yoshu*, a collection of poems compiled in the eighth century. Tsuboi Kiyotari and Tanaka Migaku, *The Historic City of Nara: An Archaeological Approach*, trans. David W. Hughes and Gina L. Barnes (Paris: UNESCO; Tokyo: Centre for East Asian Cultural Studies, 1991).

the Web site for La Tour d'Argent: http://www.latourdargent.com. See also Claude Terrail, *Ma Tour d'Argent* (Paris: Stock, 1974).

A side effect of the smoke from smudge pots: On October 14, 1947, the Los Angeles County Board of Supervisors established the first air pollution control in the United States, and probably in the world, with the Los Angeles County Air Pollution Control District. Louis C. McCabe, its director, asked Southland citrus growers to curb smoke from more than 4 million orchard heaters. A full prohibition was enacted in 1950. Sam Atwood, "The Southland's War on Smog: Fifty Years of Progress toward Clean Air," South Coast Air Quality Management District, 1997; R. L. Berg, and E. A. Wright, eds., *Frost Action and Its Control* (New York: American Society of Civil Engineers, 1984); Barry B. Coble, "Benign Weather Modification," Ph.D. dissertation, Air University, Maxwell Air Force Base, Alabama, 1996.

the broad-nosed weevil: H. N. Nigg, S. E. Simpson, D. G. Hall, L. E. Ramos, S. U. Rehman, B. Bas, and N. Cayler, "Sampling Methods as Abundance Indices for Adult *Diaprepes abbreviatus* (Coleoptera: Curculionidae) in Citrus," *Journal of Economic Entomology* 95, no. 4 (2002): 856–61; J. Garcia, " Kaolin Particle Film Knocks Out Citrus' Evil Weevil," *Agricultural Research* 49, no. 3 (2001): 18.

Aphids stand out: D. Stanley, "A Dual Citrus Threat," *Agricultural Research* 42, no. 12 (1994).

And then there is canker: J. Cabero and J. H. Graham, "Genetic Relationship among Worldwide Strains of *Xanthomonas* Causing Canker in Citrus Species and Design of New Primers for Their Identification by PCR," *Applied Environmental Microbiology* 68, no. 3 (2002): 1257–64; J. Garcia, " Tree Sentinels Help Monitor Citrus Canker," *Agricultural Research* 48, no. 12 (2000): 20–22.

the dilemma, familiar to political theorists: W. Richey, "In Florida, a Revolt against the Citrus Police," *Christian Science Monitor*, July 23, 2002; J. Gray,

"Interpreters of an Uncertain Age: Michael Oakeshott and the Political Economy of Freedom," *The World and I*, September 1988, 607.

CHAPTER 5

Max Weber's influential thesis: M. Weber, *The Protestant Ethic and the "Spirit" of Capitalism and Other Writings* (New York: Penguin USA, 2002 [1905]); R. H. Tawney, *Religion and the Rise of Capitalism* (New Brunswick, NJ: Transaction Publishers, 1998 [1926]).

Inquisitors were diligent: J. R. Magalhães, *O Algarve Economico* (Lisbon: Editorial Estampa, 1988), IV, II, 363–89. The 14th article of the 1654 Peace Treaty protected subjects of the English sovereign, who were not to be tampered with by the Inquisition on matters of religion.

the island of São Miguel: Sacuntala de Miranda, "O Ciclo da Laranja e os 'gentleman farmers' da Ilha de S. Miguel 1780–1880," doutoramento em Historia (Universidade Nova de Lisboa, 1989), 98 pp.

lines from Tennyson: Christopher Ricks, ed., *A Collection of Poems by Alfred Tennyson* (Garden City, NY: Doubleday, 1972).

the arrival of the Southern Pacific Railroad: S. E. Ambrose, *Nothing Like It in the World: The Men Who Built the Transcontinental Railroad 1865–1869* (New York: Simon & Schuster, 2000). For the contemporary railroad along the eastern seaboard, which in the 1930s would give birth to a luxury train, The Orange Blossom Special, see S. Bramson, *Speedway to Sunshine* (Erin, Ontario: Boston Mills Press, 2003); E. N. Akin, *Flagler, Rockefeller Partner and Florida Baron* (Gainesville: University Press of Florida, 1992).

the producers organized themselves into cooperatives: P. J. Dreher, "Early History of Cooperative Marketing of Citrus Fruit," *Pomona Valley Historian* 3 (1967): 139–57.

the reformulation of the Judeo-Christian utopia under Thomas Jefferson: M. Garcia, *A World of Its Own: Race, Labor, and Citrus in the Making of Greater Los Angeles, 1900–1970* (Chapel Hill: University of North Carolina Press, 2001); E. B. Black, "Ranch Life in San Antonio Canyon in the 1870s," *Pomona Valley Bulletin* 11 (1975): 47–58; L. J. Lovitt, "The Early History of Alta Loma," *Pomona Valley Historian* 8 (1972): 63–88; M. G. Stoebe, *The History of Alta Loma, California, 1880–1980* (Rancho Cucamonga: B & S Publishing, 1981); M. Stoebe and L. R. Bemis, *A History of Rialto* (Rialto, CA: Rialto Historical Society, 1999).

small religious groups of settlers: D. L. Clucas, "Transplanted Pennsylvanians in Cucamonga: The Strieby Family," *Mt. St. Antonio Historian* 15 (1979): 101–8.

the multicentric character of the greater Los Angeles area: W. B. Fulton, *The Reluctant Metropolis: The Politics of Urban Growth in Los Angeles* (Baltimore, MD: Johns Hopkins University Press, 2001); S. L. Bottles, *Los Angeles and the Automobile: The Making of the Modern City* (Berkeley: University of California

Press, 1991); R. M. Fogelson, *Downtown: Its Rise and Fall, 1880–1950* (New Haven, CT: Yale University Press, 2001).

childhood on a citrus farm: S. Straight, "House of Spirits: Remembering the Fragrant Citrus Groves of Riverside," *Westways*, September–October 2001, available at http://www.aaa-calif.com/westways/0901/sense.aspx.

George Chaffey: J. A. Alexander, *The Life of George Chaffey: A Story of Irrigation Beginnings in California and Australia* (Melbourne: Macmillan, 1928); R. E. Ellingwood, "Our Legacy: George Chaffey," *Daily Bulletin*, July 20, 1998, 12; B. Richards, "The Chaffeys: Saga of a Southern California family, Part I," *Pomona Valley Historian* 7 (1971): 25–46; Garcia, *A World of Its Own*.

The Central Valley of California: For more on the move to the Central Valley, see C. McWilliams, *Southern California: An Island on the Land* (Santa Barbara, CA/ New York: Peregrine Smith, 1973 [1946]); J. J. Parsons, "A Geographer Looks at the San Joaquin Valley," 1987 Carl Sauer Memorial Lecture, "Geography @ Berkeley," http://geography.berkeley.edu/ProjectsResources/ Publications/Parsons_SauerLect.html.

the availability of water: J. Norris Hundley, *The Great Thirst: Californians and Water—a History* (Berkeley: University of California Press, 2001).

The newer arrivals from Mexico: J. R. Garcia, *Operation Wetback: The Mass Deportation of Mexican Undocumented Workers in 1954* (Westport, CT: Greenwood Press, 1980).

Kerikeri: I took advantage of a recent (September 2002) stay for acquiring information from local residents.

CHAPTER 6

a contemporary case of scurvy: A. K. Shetty, R. W. Steele et al., "A Boy with a Limp," *Lancet* 351, no. 9097 (1998): 182.

a memoir of the journey: *Anson's Voyage around the World*, by Chaplain Richard Walter. One modern edition among many is that by The Narrative Press (Santa Barbara, CA, 2001).

A letter that Father Manuel da Nóbrega wrote home: A. Peixoto, ed., *Manuel da Nóbrega, Cartas do Brasil* (Rio de Janeiro: Academia Brasileira de Letras, 1931).

a note made by Alvaro Velho: A. Velho, *Roteiro da Primeira Viagem de Vasco da Gama (1497–1499)* (Lisbon: Agencia-Geral de Ultramar, 1969).

testimony by Thome Lopes: A. Cruz, *O Porto nas navegações e na expansão* (Lisboa: Instituto de Cultura e Lingua Portuguesa, Ministerio da Educação, 1983), 184; A. Rasteiro, *Medicina e Descobrimentos* (Coimbra: Livraria Almedina, 1992), 127.

Jesuits . . . recommended planting citrus trees: A. J. R. Russell-Wood, *Um Mundo em Movimento: Os Portugueses na Africa, Asia e America (1415–1808)* (Lisboa: Difel, 1998).

the *Encyclopédie*: This work, on the model of Chambers's *Cyclopaedia*, was a major French publication during the mid-eighteenth century. Edited by Denis

Diderot and Jean Le Rond d'Alembert, in the spirit of the Enlightenment, it was anti-Church, subversive of established dogmas, and a monument to rationality.

dissemination of plants from both India and Portugal: J. E. M. Ferrão, "Acerca da Introduçao da Laranjeira Doce em Portugal," *Anais do Instituto Superior de Agronomia* 38 (1978–1979): 197–204; *L'aventure des plantes et les découvertes portugaises* (Lisbon: Instituto de Investigaçao Cientifica Tropical, 1998); *A Aventura das Plantas e os Descobrimentos Portugueses* (Lisboa: Comissão Nacional para as Comemorações dos Descobrimentos Portugueses, 1999).

James Lind . . . proved that citrus fruit protected against scurvy: Three mistakes, common in the literature, stand corrected. James Lind's experiment was conducted during a cruise on the Atlantic Ocean, not in the South Seas; one of the "treatments" he tried was not nutmeg, but rather "the bigness of nutmeg," i.e., an amount the size of a nutmeg of the elixir whose ingredients are listed in the text. The duration of his clinical trial was two weeks, not six days. To denounce the long delay between James Lind's 1747 experiment and the decision by the Admiralty in 1795 to issue British sailors rations of lime juice is, to some extent, to engage in an anachronistic and Whig reading of the historical record. In eighteenth-century England, sailors for the Royal Navy were considered expendable. The notion of the welfare of citizens as the responsibility of the State was still inchoate. The story of fighting scurvy with citrus fruits is a case study in the progress of science: (1) common knowledge: N. Aubin, "Scorbut, ou Scurbut," in *Dictionnaire de marine, contenant les termes de la navigation et de l'architecture navale. Avec les Règles & Proportions qui doivent y être observées. Ouvrage enrichi de figures Représentant divers Vaisseaux, les principales Pièces servant à leur construction, les différens Pavillons des Nations, les Instrumens de Mathématique, Outils de Charpenterie et Menüiserie concernant la fabrique; avec les diverses fonctions des Officiers* (The Hague: Adrien Moetjens, 1742), 879, "On se peut aussi servir utilement du jus d'orange, ou de citron"; (2) discursive knowledge, as in James Lind's *Treatise*; (3) experimental inference, with an experimental test—as James Lind did in 1747. The relevant literature includes Kenneth J. Carpenter, *The History of Scurvy and Vitamin C* (Cambridge: Cambridge University Press, 1986); F. E. Cuppage et al., "James Cook's Eighteenth-Century Prevention of Scurvy by the Use of Indigenous Plants as Dietary Supplements," *Terrae Incognitae: The Journal for the History of Discoveries* 26 (1994): 37–45; Christopher Lloyd, ed., *The Health of Seamen* ([London]: The Navy Records Society, 1965); D. P. Thomas, "Sailors, Scurvy and Science," *Journal of the Royal Society of Medicine* 90 (1997): 50–54; C. Lloyd, "The Introduction of Lemon Juice as a Cure for Scurvy," *Bulletin of the History of Medicine* 35 (1961): 123–32; D. P. Thomas, "Experiment versus Authority: James Lind and Benjamin Rush," *New England Journal of Medicine* 281 (1969): 932–34; D. W. Amory, "Lind, Scott, Amundsen and Scurvy," *Journal of the Royal Society of Medicine* 90 (1997): 299; J. H. Baron, "Scurvy, Lancaster, Lind, Scott and Almroth

Wright," *Journal of the Royal Society of Medicine* 90 (1997): 415; D. Baxby, "Lind's Clinical Trial and the Control of Scurvy," *Journal of the Royal Society of Medicine* 90 (1997): 526–27; E. M. Bardolph and R. H. Taylor, "Sailors, Scurvy, Science and Authority," *Journal of the Royal Society of Medicine* 90 (1997): 238; R. Bartlett, "Britain, Russia, and Scurvy in the Eighteenth Century," *Oxford Slavonic Papers* 29 (1996): 23–43; R. E. Hughes, "James Lind and the Cure of Scurvy: An Experimental Approach," *Medical History* 19 (1975): 342–51; J. Glass, "James Lind, M.D., Eighteenth Century Naval Medical Hygienist: Biographical Notes with an Appreciation of the Naval Background," *Journal of the Royal Navy Medical Service* 35 (1949): 1–20, 68–86; A. P. Meiklejohn, "The Curious Obscurity of Dr. James Lind," *Journal of the History of Medicine and Allied Sciences* 9 (1954): 304–10; C. Lawrence, "Scurvy, Lemon Juice and Naval Discipline 1750–1815," in *The Second National Medical History Conference*, Perth, Western Australia, 1991, The Society and University of Melbourne Medical History Unit, *Occasional Papers on Medical History* 5 (1991): 227–32.

in spite of the efforts of many scientists: George Budd (1808–1882), Professor of Medicine at King's College, London, published in 1842, in the *London Medical Gazette*, a series of articles entitled "Disorders Resulting from Defective Nutriment." Three of the diseases described by Budd were, as we now know, deficiencies of vitamins A, C, and D. He believed that scurvy was due to the "lack of an essential element which is hardly too sanguine to state will be discovered by organic chemistry or the experiments of physiologists in a not too distant future." In 1907, A. Holst and T. Fröhlich of Norway induced scurvy in guinea pigs (animals also subject to deficiencies of vitamin C) by dietary means. E. G. Hopkins in 1912 demonstrated the presence of essential growth factors in milk. In 1917, H. Chick and M. Hume reported the antiscurvy factor's distribution in a number of foodstuffs. The following year, A. Harden and S. S. Zilva, of the Lister Institute, London, published their fractionation study on lemon juice. In 1932, W. W. Waugh and Charles G. King published "The Isolation and Identification of Vitamin C" with a photograph of vitamin C crystals. K. F. Kiple and K. C. Ornelas, eds., *The Cambridge World History of Food* (Cambridge: Cambridge University Press, 2000), 755–66.

citric acid cycle: L. Stryer, *Biochemistry* (San Francisco: W. H. Freeman, 1995).

a Dutch professor: Louis Rosenfeld, "Vitamine—Vitamin: The Early Years of Discovery," *Clinical Chemistry* 43 (1997): 680–85.

to name his crystalline sample "ignose": *The Cambridge World History of Food*, 756.

smelled of the kitchen: Szent-Györgyi and other scientists spurned the topic of nutrition, associated for them with "women's work." "For some inexplicable, childish reason," he later said, "I felt that vitamins were a problem for the cook." R. W. Moss, *Free Radical: Albert Szent-Györgyi and the Battle over Vitamin C* (New York: Paragon House, 1988) 76.

Americans were eager: Moss, *Free Radical*, 77.

Charles Glen King (1896–1988): M. Asbell, "Finding Aid to the Charles Glen King Papers, 1918–1988," National Library of Medicine, MS C 473: Biographical note.

went to Hungary: A. Szent-Györgyi, "Oxidation, Energy Transfer, and Vitamins" (Nobel Lecture, 1937).

told his Hungarian supervisor: Moss, *Free Radical*, 79.

King wrote to Svirbely: Moss, *Free Radical*, 80.

both enthusiastic and trusting: for one affectionate portrait of Albert Szent-Györgyi (among many), see James D. Watson, *Genes, Girls and Gamow: After the Double Helix* (New York: Alfred A. Knopf, 2002).

Szent-Györgyi was incensed: Moss, *Free Radical*, 81–85. King's behavior, if indeed he stole some results from Szeged, has to be understood in the context of his having invested many years of labor in the search for the anti-scurvy factor. He must have seen Szent-Györgyi as an opportunistic upstart.

finally granted U.S. patent no. 2,233,417: Moss, *Free Radical*, 279.

!Kung subsistence diet: R. B. Lee and I. D. Vore, eds., *Kalahari Hunter-Gatherers: Studies of the !Kung San and Their Neighbors* (Cambridge MA: Harvard University Press, 1976); R. B. Lee, *The !Kung San: Men and Women in a Foraging Society* (Cambridge: Cambridge University Press, 1979).

Take the Inuit: V. Stefansson, "Food and Food Habits in Alaska and Northern Canada," in *Human Nutrition, Historic and Scientific*, ed. I. Galdston (New York: International University Press, Inc., 1960), 23–60; G. Palsson, ed., *Writing on Ice: The Ethnographic Notebooks of Vilhjalmur Stefansson* (Hanover, NH: University Press of New England, 2001); H. Brody, *The Other Side of Eden: Hunters, Farmers, and the Shaping of the World* (New York: North Point Press [Farrar, Straus & Giroux], 2001).

a preventive and cure for the common cold: L. Pauling, *Vitamin C and the Common Cold* (San Francisco: W. H. Freeman, 1970); "Ascorbic Acid and the Common Cold," *American Journal of Clinical Nutrition* 24 (1971): 1294–99; "The Significance of the Evidence about Ascorbic Acid and the Common Cold," *Proceedings of the National Academy of Sciences USA* 68 (1971): 2678–81; "Are Recommended Daily Allowances for Vitamin C Adequate?" *Proceedings of the National Academy of Sciences USA* 71 (1974): 4442–46.

a cancer fighter: E. Cameron and L. Pauling, *Cancer and Vitamin C: A Discussion of the Nature, Causes, Prevention, and Treatment of Cancer With Special Reference to the Value of Vitamin C* (Philadelphia: Camino Books, 1993).

a climb in the Alps: The classic books, at least in the French language, on climbing the Alps on skis are P. Traynard and C. Traynard, *Alpes et neige—101 sommets à ski* (Grenoble: Arthaud, 1965); and C. Traynard and P. Traynard, *Ski de Montagne* (Grenoble: Arthaud, 1974).

Perrier citron tranche: The Perrier company had an advertising genius, Jean Davray, to boost and finesse its sales from 1946 until his death in 1985. His advertising campaigns have become classics taught in design schools. He recruited the best poster artists: Jean Effel, Morvan, Savignac, Villemot, Jean Carlu, even Salvador Dali. He made Perrier into a sponsor for the Tour de France bicycle race. He recruited the writer Colette, who wrote of Perrier, "*Une eau qui bondit quand on la débouche. Une eau qui rit. Une eau qui est dans la bouche comme une poignée d'aiguilles."* (A water that springs when the cap is removed. A water that laughs. A water that stands in the mouth like a handful of needles.) J. Watin-Augouard, "Histoire d'une marque: Perrier," *La revue des marques* 12 (1995); D. Cauzard, J. Perret, Y. Ronin, and V. Mitteau, *Le livre des Marques* (Paris: Edition du May, 1993).

CHAPTER 7

Lasker's innovation: C. A. Goodrum and H. Dalrymple, *Advertising in America: The First 200 Years* (New York: Harry N. Abrams, 1990); E. Applegate, *The Ad Men and Women: A Biographical Dictionary of Advertising* (Westport, CT: Greenwood Press, 1994); A. D. Lasker, "The Personal Reminiscences of Albert Lasker," *American Heritage*, December 1954, 77; A. D. Lasker, *The Lasker Story . . . As He Told It* (Chicago: Advertising Publications, 1963).

pasteurization: Most American urban milk supplies (90 percent) were pasteurized by 1920. Exposure to tuberculosis through contaminated milk remained a problem, particularly for infants. Orange juice started being pasteurized in the 1920s in both California and Florida.

an integral part of the American breakfast: An "English breakfast" may include cereal with milk, dry or cooked (porridge); boiled or fried eggs; cooked or fried meats (ham, bacon, sausages, patties, steak); toast with butter, jam, or marmalade; stewed fruit (prunes); and kippers. J. Burnett, *Liquid Pleasures: A Social History of Drinks in Modern Britain* (London: Routledge, 1999).

The California Fruit Growers Exchange: The CFGE was an offshoot of the earlier Southern California Fruit Exchange (SCFE), a cooperative formed and incorporated by growers in 1895. The CFGE operated at an at-cost basis and enjoyed the legal power of a stock corporation. In 1913, when Powell took the helm, the CFGE sold 30,000 carloads of fruit from 6,000 growers. By 1922, it shipped 55,000 carloads of fruit per year.

headed by G. Harold Powell: In 1904, Powell, then with the USDA Bureau of Plant Industry and a protégé of Liberty Hyde Bailey, quelled an uproar among California orange growers, who were complaining that 25 percent of their shipments were decaying en route to New York and other markets in the East. In 1910, at the urging of the growers, Powell became executive secretary of the newly formed California Citrus Protective League (CPL). In 1913, Powell was appointed general manager of the CFGE, which had existed since 1905.

the structure of a modern corporation: H. V. Moses, "G. Harold Powell and the Corporate Consolidation of the Modern Citrus Enterprise, 1904–1922," *Business History Review* 69, no. 2 (1995): 119.

one of the most impassioned passages: J. Steinbeck, *The Grapes of Wrath*, in *The Grapes of Wrath and Other Writings 1936–1941* (New York: Library of America, 1996), chap. 25, 578–80.

50 percent more oranges than Florida: This relative abundance of fruit led to an effort by the citrus industry to displace apple preparations, plums, and prunes on the American breakfast table. Central was a "drink your fruit" advertising campaign: H. J. Stover, "The Manufacture and Use of California Canned Orange Juice" (Berkeley: University of California, College of Agriculture, Agricultural Experimental Station, 1936). Approximately 50 percent of the fruit weight is juice.

Some of these chemicals can interact with drugs: P. Mitchell, "Grapefruit Juice Found to Cause Havoc with Drug Uptake," *Lancet* 353, no. 9161 (1999): 1335–36.

About twenty-five different chemicals: R. Coxeter, "Orange Juice—Flavor and Odor," *Business Week*, December 6, 1993, 85.

Frozen concentrated orange juice (FCOJ): R. F. Matthews, "Frozen Concentrated Orange Juice from Florida Oranges," University of Florida Cooperative Extension Service, Food Science and Human Nutrition, Institute of Food and Agricultural Sciences, April 1994, http://edis.ifas.ufl.edu/CH095.

An orange appeals to the taste: X. M. Gao and J.-Y. Lee, "An Application of a Multiple Cause Variable Model for Consumer Perception of Orange Juice," *Applied Economics* 25 (1993): 207–12. A good summary of the early literature is K. S. Keeley and J. E. Kinsella, "Orange Juice Quality with an Emphasis on Flavor Components," *CRC Critical Reviews in Food Science and Nutrition* 11, no. 1 (1978): 1–40. Among more recent studies, see K. L. Goodner, P. Jella, and R. L. Rouseff, "Determination of Vanillin in Orange, Grapefruit, Tangerine, Lemon And Lime Juices Using GC-Olfactometry and GC-MS/MS," *Journal of Agricultural and Food Chemistry* 48, no. 7 (2000): 2882–86; A. Buettner and P. Schieberle, "Evaluation of Aroma Differences between Hand-Squeezed Juices from Valencia Late and Navel Oranges by Quantitation of Key Odorants and Flavor Reconstitution Experiments," *Journal of Agricultural and Food Chemistry* 49, no. 5 (2001): 2387–94. For similar work on grapefruit juice, see A. Buettner and P. Schieberle, "Characterization of the Most Odor-Active Volatiles in Fresh, Hand-Squeezed Juice of Grapefruit (*Citrus paradisi* Macfayden)," *Journal of Agricultural and Food Chemistry* 47, no. 12 (1999): 5189–93; A. Buettner and P. Schieberle, "Evaluation of Key Aroma Compounds in Fresh, Hand-Squeezed Juice of Grapefruit (*Citrus paradisi* Macfayden) by Quantitation and Flavor Reconstitution Experiments," *Journal of Agricultural and Food Chemistry* 49, no. 3 (2001): 1358–63; J. Lin, R. L. Rouseff, S. Barros, and M. Naim, "Aroma Composition Changes in Early Season Grapefruit Juice Produced from Thermal Concentration," *Journal of Agricultural and Food Chemistry* 50, no. 4 (2002): 813–19.

Arnold O. Beckman: A. Thackray and J. Minor Myers, *Arnold O. Beckman: One Hundred Years of Excellence* (Philadelphia: Chemical Heritage Foundation, 2000); G. B. Kauffman and L. M. Kauffman, "A Legend in His Own Time: Arnold Beckman's Scientific and Technological Achievements Ushered In the instrumentation revolution," *Chemist* 10, October (2001): 63–64; Beckman, A. O., and H. E. Fracker. "Apparatus for Testing Acidity," U.S. Patent 2,058,761, 1936.

fraudulent juice: Paula Kurtzweil, "Fake Food Fight," *FDA Consumer Magazine*, March–April, 1999, available at http://www.fda.gov/fdac/features/1999/299_food.html; N. H. Low, A. Brause, and E. J. Wilhelmsen, "Normative Data for Commercial Pineapple Juice from Concentrate," *Journal of AOAC International* 77 (1994): 965–75; N. H. Low, "[On Detection of Adulterated Fruit Juices by Gas Chromatography]," *Fruit Processing* 11 (1995): 362–67; N. H. Low and D. A. Hammond, "[Detection by GC of Adulterated Fruit Juices]," *Fruit Processing* 4 (1996): 135–39; N. H. Low, ""Determination of Fruit Juice Authenticity by Capillary Gas Chromatography with Flame Ionization Detection," *Journal of AOAC International* 79 (1996): 724–37.

José Cutrale, Jr.: R. Cohen, "Citrus King: Brazil's Jose Cutrale, Helped by Coca-Cola, Is Taking On Florida—His Orange Juice Concentrate Is Sold to Minute Maid; U.S. Growers Cry Foul—Glut and Saturated Markets," *Wall Street Journal*, January 22, 1987, 1; J. L. d. Souza, "Os frutos da competência," *Epoca*, June 28, 1999, 58.

pushing up the price of orange juice: Weather is one of the parameters affecting the price of orange juice: R. Roll, "Orange Juice and Weather," *American Economic Review* 74, no. 5 (1984): 861–80. When the orange groves freeze, turning the fruit into juice is the only option for the growers, who suffer huge economic damage. For instance, the California Farm Bureau Federation in Sacramento estimated the 1998 freeze's damage to the state's citrus crops at $591 million. Growers estimated that 50 percent to 75 percent of their navel orange crops were lost. More than 80 percent of the state's orange crops were produced in the freeze-hit counties of Tulane, Fresno, and Kern. Prices for concentrated orange juice futures settled on December 28, 1998, at $1.06 a pound, affected by a smaller than expected Florida crop: the 1998–1999 crop was predicted to be 190 million 90-pound boxes of oranges, down from 1997–1998's production of 244 million boxes. T. Ewing and S. Kravetz, "Freeze May Squeeze Prices Up and Down—Sale of Oranges Damaged in Hard-Hit California Could Cut Cost of Juice," *Wall Street Journal*, December 29, 1998, A2. Not long ago, growers were still uninsured against the risk of citrus crop losses, despite their recurring occurrence. At the time of the 1990 freeze, insurance covered a small fraction of California's multimillion-dollar citrus crop loss triggered by a pre-Christmas freeze. Most commercial growers had not protected their irrigation systems, which sustained widespread damage. Citrus growers sought federal disaster relief for the crop loss, estimated at more than $350 million: L. Kertesz and J. Wojcik, "Growers Uninsured for Citrus Crop Loss," *Business Insurance* 24, no. 53 (1990): 1–2.

Swiss immigration to Brazil: G. Seyferth, "Imigração e colonização alemã no Brasil: Uma revista da bibliografia," *Boletim Informativo e Bibliografico de Ciencias Sociais* 25 (1988): 3–55; G. Ducotterd and R. Loup, *Terre! Terre!*

Récit historique de l'émigration suisse au Brésil en 1819 (Fribourg, Switzerland: Renaissance Rurale, 1939); M. Nicoulin, *La Genèse de Nova Friburgo* (Fribourg, Switzerland: Editions Universitaires, 1981).

Avocado with Lime Juice: Is hydrolysis of the avocado fruit by the lime juice due to the acidity only, or to an enzyme present in the lime? On the avocado as foodstuff, see J. Meadows, "Florida Food Fare: Avocado," *Sarasota Herald-Tribune*, November 19, 1998, Food Section. On the presence of vanillin in citrus juices, see Goodner et al., "Determination of Vanillin." On phenolic antioxidants, see M. Berhow, B. Tisserat, K. Kanes, and C. Vandercook, "Survey of Phenolic Compounds Produced in Citrus," USDA Technical Bulletin no. 1856 (Peoria, IL: USDA Agricultural Research Service, 1998). On how the ideal perceived sweetness of a food is related to sex, age, and culture, see M. T. Conner and D. A. Booth, "Preferred Sweetness of a Lime Drink and Preference for Sweet over Non-sweet Foods, Related to Sex and Reported Age and Body Weight," *Appetite* 10, no. 1 (1988): 25–35.

on the *caïpirinha*: P. A. Harrison, *Behaving Brazilian: A Comparison of Brazilian and North American Social Behavior* (Rowley, MA: Newbury House Publishers, 1983); J. Oliver, *The Naked Chef Takes Off* (New York: Hyperion Books, 2001).

creatures of habit when it comes to food: For most immigrants to the United States, this was due in part to the limited range of foods imposed on them by poverty in their home countries. H. R. Diner, *Hungering for America: Italian, Irish, and Jewish Foodways in the Age of Migration* (Cambridge, MA: Harvard University Press, 2002).

the preventive effect of orange juice . . . against stroke: K. J. Joshipura, A. Ascherio, J. E. Manson, M. J. Stampfer, E. B. Rimm, F. E. Speizer, C. H. Hennekens, D. Spiegelman, and W. C. Willett, "Fruit and Vegetable Intake in Relation to Risk of Ischemic Stroke," *JAMA* 282, no. 13 (October 6, 1999): 1233–39. The ensuing PR blitz by Florida Citrus is described in D. Campbell, "Selling It!" *Rural Cooperatives* 5 (2001), September–October, available at http://www.rurdev.usda.gov/rbs/pub/sep01/selling.htm. Some took issue with this campaign, claiming overkill: Barrett, Stephen, "Misleading Ads from the Florida Department of Citrus" (2000), http://www.nutriwatch.org/08Ads/citrus.html.

secret ink: The late Professor Georges Bram, of the Institute for Molecular Science, Université Paris-Sud, Orsay, experimented with lemon juice as "invisible" ink. He made aqueous solutions of ascorbic acid and citric acid. Once revealed by heat, ascorbic acid leaves a darker trace than citric acid, more closely resembling that from lemon juice itself. Professor Bram ascribed this effect, quite reasonably, to thermal decomposition of the organic material dissolved in water. Any colorless juice from an organic source, animal or plant, may show similar effects. Hervé This, "Faisons des expériences simples: La culture scientifique, un enjeu de la Gastronomie moléculaire," available at Le lycée Chateaubriand, http://www.lycee-chateaubriand.fr/cru-atala/publications/this.htm.

which he allegedly invented: While the name implies that the painter Carpaccio invented this dish, this story is too good to be true. In fact, Giuseppe Cipriani, the chef and owner of Harry's Bar—a renowned watering hole in Venice founded in 1931—devised the carpaccio in 1951; see http://www.cipriani.com/cipriani/Consigli/carniricettee.htm.

CHAPTER 8

an otherwise healthy New York City bartender: A. C. Cardallo, A. M. Ruszkowski, and V. A. DeLeo, "Allergic Contact Dermatitis Resulting from Sensitivity to Citrus Peel, Geraniol, And Citral," *Journal of the American Academy of Dermatology* 21, no. 2 (1989): 395–97; M. Matura, A. Goossens, O. Bordalo, B. Garcia-Bravo, K. Magnusson, K. Wrangsjo, and A. T. Karlberg, "Oxidized Citrus Oil (R-limonene): A Frequent Skin Sensitizer in Europe," *Journal of the American Academy of Dermatology* 47, no. 5 (2002): 709–14.

Nonsensory evidence: Identification of the substances from the essential oils of citrus peel is routine: B. Steuer, H. Schulz, and E. Läger, "Classification and Analysis of Citrus Oils by NIR Spectroscopy," *Food Chemistry* 72, no. 1 (2001): 113–17. Mass spectrometry is an even more powerful technique than near infrared (NIR) spectroscopy, applied, for instance, to the *yuzu* essential oils: M. Sawamura, T. Ito, A. Une, H. Ukeda, and Y. Yamasaki, "Isotope Ratio by HR GC-MS of Citrus *Junos tanaka* (*yuzu*) Essential Oils: m/z 137/136 of Terpene Hydrocarbons," *Bioscience, Biotechnology, and Biochemistry* 65, no. 12 (2001): 2622–29.

adapts to the plant's circumstances: D. Sun and P. D. Petracek, "Grapefruit Gland Oil Composition Is Affected by Wax Application, Storage Temperature, and Storage Time," *Journal of Agricultural and Food Chemistry* 47, no. 5 (1999): 2067–69; S. A. Vekiari, E. E. Protopapadakis, P. Papadopoulou, D. Papanicolaou, C. Panou, and M. Vamvakias, "Composition and Seasonal Variation of the Essential Oil from Leaves and Peel of a Cretan Lemon Variety," *Journal of Agricultural and Food Chemistry* 50, no. 1 (2002): 147–53.

there are treasures within citrus peel: J. Lawless, *The Illustrated Encyclopedia of Essential Oils* (Rockport, MA: Element Books, 1995); E. Guenther, *The Essential Oils* (New York: Van Nostrand, 1948).

The size of the gland's opening: I. Panchev and S. Karageorgiev, "Investigation of Some Physical Characteristics of Plant Structures Which Are Used as Sources of Pectic Substances," *International Journal of Food Science & Technology* 35, no. 3 (2000): 341–51.

protection against insects: The presence of terpenes in the peel is an evolutionary advantage: see, inter alia, F. C. Ezeonu, G. I. Chidume, and S. C. Udedi, "Insecticidal Properties of Volatile Extracts of Orange Peel," *Bioresource Technology* 76, no. 3 (2001): 273–74. A seabird uses similar molecules for its chemical defense against parasites: H. D. Douglas III, J. E. Co, T. H. Jones, and W. E. Conner, "Heteropteran Chemical Repellents Identified in the Citrus Odor

of a Seabird (Crested Auklet: *Aethia cristatella*): Evolutionary Convergence
in Chemical Ecology," *Naturwissenschaften* 88, no. 8 (2001): 330–32. A substance from orange peel is an attractor to insect predators of Colorado potato
beetles, the *doryphores* that snatched away from our hungry mouths the coveted potatoes grown in our garden at the end of World War II (see chap. 1): A.
Coghlan, "Double-Crossed," *New Scientist* 2226 (2000): 14–17.

perfumed dust to enhance a dish: L. M. Ohr, "A Foray into Flavors," *Prepared
Foods*, 302 (March), 2002. Among new products introduced in the United
States in 2001, there were 152 orange-flavored beverages, an increase of 121
percent over the previous year.

various tools devised for this purpose: K. Farrell-Kingsley and J. Charatan, "Lemon
Aids," *Vegetarian Times*, August 2000.

which he named hesperidin: Lebreton's original publication shows experimental
observation at its best: M. Lebreton, "Sur la matière cristalline des orangettes,
et analyse de ces fruits non encore développés, famille des Hespéridées,"
Bulletin des travaux de la société de pharmacie VIII, Juillet (1828): 377–92.

This is even more true of citral: Among the terpenes in citrus peel, citral is of the
utmost economic importance: A. Floreno, "Citral," *Chemical Marketing Reporter* 15 (1995): 18. Vulnerable as it is to weather and other vagaries, the
orange oil market is very volatile (!): D. Mazzaro, "Orange Oil, D-Limonene
Market Unsettled Due to Brazilian Delays," *Chemical Marketing Reporter* 258,
no. 4 (2000): 18.

they provide the aroma with its signature: N. T. M. Tu, V. Onishi, H. S. Choi, Y.
Kondo, S. M. Bassore, H. Ukeda, and M. Sawamura, "Characteristic Odor
Components of *Citrus sphaerocarpa* Tanaka (Kabosu) Cold-Pressed Peel Oil,"
Journal of Agricultural and Food Chemistry 50, no. 10 (2002); H. S. Choi, Y.
Kondo, and M. Sawamura, "Characterization of the Odor-Active Volatiles in
Citrus Hyuganatsu (*Citrus tamurana* Hort. ex Tanaka)," *Journal of Agricultural
and Food Chemistry* 49, no. 5 (2001): 2404–8; Vekiari et al., "Composition and
Seasonal Variation."

soothing . . . smell: Not unlike listening to Mozart, it reduces stress at the dentist's.
J. Lehrner, C. Eckersberger, P. Walla, G. Potsch, and L. Deecke, "Ambient
Odor of Orange in a Dental Office Reduces Anxiety and Improves Mood in
Female Patients," *Physiology and Behavior* 71, no. 1–2 (2000): 83–86.

Ben Jonson: *Christmas, His Masque*, first performed in December 1616, first published in 1640.

a major source of raw material for perfumery: B. Meyer-Warnod, "Natural Essential Oils: Extraction Processes and Applications to Some Major Oils," *Perfumer
& Flavorist* 9 (1984): 93.

CHAPTER 9

the Atalanta myth: Consult any standard Greek mythology. Considered jointly
with the related myth (through the golden apples) of the garden of the

Hesperides, it amounts to a cosmology; to be more precise, an account of the circular motion of the Sun, perhaps solar eclipses: A. Ballabriga, "Variations sur la coïncidence des opposés aux confins occidentaux et orientaux de l'univers," in *Le Soleil et le Tartare: L'image mythique du monde en Grèce archaïque* (Paris: Editions de l'Ecole des Hautes Etudes en Sciences Sociales, 1986), 75–106.

Atalanta Fugiens: This enigmatic sequence of emblems, entitled *Atalanta Fugiens, hoc est, Emblemata Nova De Secretis Chymica, Accommodate partim oculis & intellectui, figures cupro incises, adjestisque sententiis, Epigrammatis & notis, parim auribus & recreationi animi plus minus 50 Fugis Musicalibus trium Vocum, . . .* (Oppenheim: Johann Theodor de Bry, 1617) has had modern editions; an English translation appeared in 1989, produced by Phanes Press of Grand Rapids, MI.

Sandro Botticelli: A good discussion of *Primavera* is F. Zöllner, "Zu den Quellen und zur Ikonographie von Sandro Botticellis 'Primavera,'" *Wiener Jahrbuch für Kunstgeschichte* 50 (1997): S. 131–57; see also F. Hart, *Sandro Botticelli* (New York: Harry N. Abrams, 1953); K. Clark, *The Drawings by Sandro Botticelli for Dante's* Divine Comedy: *After the Originals in the Berlin Museums and the Vatican* (New York: Harper and Row, 1976).

War of the Oranges: J. Godechot, "Le Portugal et la Révolution," in *Arquivos de Centro cultural Portugues VII* (1973): 279–97; J.-F. Labourdette, *Histoire du Portugal* (Paris: Fayard, 2000); N. Gotteri, "Napoléon et le Portugal," Institut Napoléon (Paris La Sorbonne), 6 avril 2002, http://perso.wanadoo.fr/Cerclouisdenar/on_line/Napoleon_et_le_Portugal.htm; A. Ventura, *O Combate de Flor da Rosa (Conflito Luso-Espanhol de 1801)* (Lisbon: Ediçoes Colibri, 1996).

"excite by all possible means Spain to go to war against Portugal": Thierry Lentz, "Les relations franco-espagnoles. Réflexions sur l'avant-guerre 1789–1808," *Revue du Souvenir Napoleônien*, no. 399 (1995), 4–20; see http://www.napoleon.org/fr/salle_lecture/articles/files/relations_franco-espagnoles_Reflexions_sur.asp.

"Godoy allows himself acts of violence": Lentz, "Les relations franco-espagnoles."

The small town of Binche: J. Delmelle, *Binche: La cité des gilles* (La Madeleine-lez-Lille: Editions Actica, 1972); S. Glotz, *De Marie de Hongrie aux Gilles de Binche: Une double réalité historique et mythique* (Bruxelles: Wallonie-Bruxelles, 1995). A. Garin, *Binche et le carnaval: Binche, cité impériale, son histoire, son folklore, ses richesses et ses traditions* (Charleroi: Imprimerie provinciale du Hainaut, 1998) makes the connection with Bacchanals and Lupercals of Roman antiquity.

the so-called Gilles: E. Matthieu, *Quelques mots sur l'origine des Gilles* (Enghien: Imprimerie A. Spinet, 1899).

Inca costumes: A legend explaining the origin of these costumes, documented only in accounts by journalists at the end of the nineteenth century, goes back to a seven-day festival occasioned in Binche by the visit of Charles V, Roman Germanic Emperor, on August 22, 1549, which Mary of

Hungary threw in his honor. Did Incas attend, real Incas or revelers costumed as Incas? See J. Huynen, *La Mascarade sacrée: Binche témoigne* (Brussels: Louis Musin, 1979).

The pelted oranges in the Binche Carnival: S. Glotz, *Le carnaval de Binche* (Bruxelles: Editions du Folklore brabançon, n.d.).

A traveler wrote about Carnival in Rome in 1847: C. G. Leland, "The Carnival at Rome—1847," *Godey's Lady's Book*, February 1850, http://www.history. rochester.edu/godeys/02-50.htm.

Carnival in Ivrea: "Ivrea: The History of the Carnival," http://www. carnevalediivrea.com.

a festival celebrating the return of light: M. Eliade, *Traité d'histoire des religions* (Paris & Lausanne: Payot, 1975), 342.

they accompanied festive occasions: E. C. Gaskell, *Mary Barton* (London: Everyman's Library, 1965), 72; E. C. Gaskell, *North and South* (London: Everyman's Library, 1967), 73. In Greece, completion of a roof is celebrated by the attachment of a devotional cross with oranges nailed at the extremities: Eleni N. Gage, *North of Ithaka* (New York: St. Martin's Press, 2005).

orange vendors in Palermo: M. Amari, *Storia dei musulmani di Scilia* (Catania: Dafni, 1986), vol. 3, pt. 5, 920, n. 5.

orange vendors in Cairo: E. W. Lane, *The Manners and Customs of the Modern Egyptians* (London and Toronto: J. M. Dent; New York: E. P. Dutton, 1908), 327.

at the Globe Theatre: A. Gurr, *Playgoing in Shakespeare's London* (Cambridge: Cambridge University Press, 1987): "Hazelnuts were the most popular theatre snack. Once the patrons had secured a spot, they could enjoy talking loudly with neighbors, smoking, etc. while the performance was in progress. They could refresh themselves from the heat by purchasing treats from vendors offering beer, water, oranges, nuts, gingerbread, and apples, all of which were occasionally thrown at the actors." When in *Much Ado About Nothing*, a character said: "There, Leonato, take her back again: / Give not this rotten orange to your friend; / She's but the sign and semblance of her honor" (act 4, scene 1, lines 31–33), the insult had a familiar tone to the audience.

such cities as York and Manchester: Other British cities likewise associated oranges with theatergoing: "At the theatre, gingerbread and oranges were the traditional fare, the thirteen-year old Mary Worsley paying 'for oranges 6d; for seeing a play 2d' on 2 June 1697. Some of these items could be bought as 'fast food' from the street traders, such as Mary Atkinson the orange-woman in 1713, and her successor, the orange-man who is seen crying his 'Sweet China Oranges' in James Kendrew's *Cries of York* printed in the opening years of the nineteenth century." E. White, ed., *York and Manchester: Feeding a City: York. The Provision of Food from Roman Times to the Beginning of the Twentieth Century* (Allaleigh House, Blackawton, Tornes, Devon: Prospect Books, 1993), 163. "From abroad, the only fruits that reached Manchester in any quantity were apples and oranges . . . The Liverpool and Manchester Railway was bringing sufficient quantity [of oranges] to Manchester in 1831–2 for the company to think it worthwhile to adopt special precautions against persistent pilferage. Strangely, this appears to have been a particular problem with consignments of oranges and does seem to indicate that the individual orange was reckoned a price worth stealing. Yet a witness to the 1839 Select Committee argued that they could no longer be considered luxuries, and they were widely sold on the streets of London." Roger Scola, *Feeding the Victorian City. The Food Supply of Manchester, 1770–1870*, ed. W. A. Armstrong and Pauline Scola (Manchester: Manchester University Press, 1992), 123.

incident . . . on December 20, 1700: "John Cowland, Gentleman," in *The Complete Newgate Calendar* (London: Navarre Society, 1926). The orange-women were routinely insulted by gentlemen, as the language in a review of a contemporary play makes obvious: "I will take for granted, that a fine Gentleman should be honest in his Actions, and refined in his Language. Instead of this, our Hero in this Piece [Sir George Etheredge's *The Man of Mode, or Sir Fopling Flutter*] is a direct Knave in his Designs, and a Clown in his Language. . . . As to his fine Language; he calls the Orange-Woman, who, it seems, is inclined to grow fat, *An Over-grown Jade*, with a flasket of Guts before her; and salutes her with a pretty Phrase of *How now, Double Tripe!*" *The Spectator*, May 15, 1711, available at http://meta.montclair.edu/spectator/text/may1711/no65.html. The text of the play is available at http://ourcivilisation.com/smartboard/shop/etherege/index.htm.

Nell Gwynn: S. Coote, *Royal Survivor: The Life of Charles II* (London: Hodder and Stoughton, 2001).

my father and I: We traveled from Paris to Rio de Janeiro, with refueling stops in Madrid, Dakar, and Recife on a four-propeller Lockheed L-749 Constellation plane, which held about a hundred passengers. It was a remarkable machine. When we took off from Dakar, one of the engines caught fire. We went back, the repair was made, and we were aloft again a couple of hours later. B. Buck's *North Star Over My Shoulder* (New York: Simon & Schuster, 2002) has similar tall-sounding stories, upholding my recollection. We arrived about July 20, 1950. The final of the World Cup, Coupe Jules Rimet, had been held on July 16 in the Maracana Stadium. More than 200,000 spectators witnessed, at the 79th minute, Gigghia scoring the second goal for Uruguay, which defeated Brazil 2–1. The Hotel Excelsior no longer exists in its location in Central Rio, near Avenida Rio Branco. There is a hotel by that name, but it is located on Copacabana nowadays.

the luxurious sensuality sometimes attributed to Afro-Brazilians: The famous Brazilian novelist Jorge Amado (1912–2001), whose books are predominantly set in the land of the *coronels*, the Brazilian Northeast, devoted one of his most popular novels to the character of a mulatress, epitomizing the Brazilian stereotypes about *a mulata*: J. Amado, *Gabriela Cravo e Canela* [Gabriela Clove and Cinnamon] (Bard Books, 1998 [1958]). An interesting article discusses the literary status of the mulatress in the novel *Clara dos Anjos*, published posthumously in 1948 by Lima Barreto (1881–1922):

C. H. Gileno, "*Clara dos Anjos*: uma reflexão sobre o status da mulata no Brasil do inicio do seculo XX," *Ciência & Tropico, Recife* 29, no. 1 (2001): 125–46. On the general topic of the mixed-blood woman in Brazilian culture, see Christiane Izard, "La femme de couleur dans l'histoire de l'Amérique latine et ses images littéraires," Thèse de 3e cycle/Etudes latino-américaines, dir. Jeanine Potelet, Université Paris X-Nanterre, 1987. Carmen Miranda, another example of the stereotype exported from Brazil, had a highly successful American career: J. Dibbell, "Notes on Carmen," *Village Voice*, October 29, 1991; M. Gil-Montero, *Brazilian Bombshell: The Biography of Carmen Miranda* (New York: D. I. Fine, 1989). Among the resonance evoked by the Brazilian mulatress's sensuality, I select for attention a book of erotic poems, in the Brazilian lineage of Mario da Andrade, Oswald's *Mon envie, ma soeur* (Paris: Editions 00h00, 2000).

a casa-grande: On life in the big house, see G. Freyre, *The Masters and the Slaves* [*Casa-Grande & Senzala*]: *A Study in the Development of Brazilian Civilization* (Berkeley: University of California Press, 1987); G. Freyre, *The Mansions and the Shanties* [*Sobrados e Mucambos*]: *The Making of Modern Brazil* (Westport, CT: Greenwood, 1980); G. Freyre, ed., *Livro do Nordeste* (Recife: Officio do diario de Pernambuco, 1925).

Mediterranean cuisine: E. David, *A Book of Mediterranean Food* (New York: New York Review of Books, 2002 [1950]).

The Atlantic Ocean is definitely not a member of the family: The masterpiece song by Dorival Caymmi, "O Mar," evokes the plight of the *jangadeiros*: D. Caymmi, *Trinta Sucessos de Dorival Caymmi* (Rio de Janeiro: Magione, Filhos & Cia Ltda, 1949).

Brazilian rituals . . . imported from Africa: R. Bastide, *The African Religions of Brazil: Toward a Sociology of the Interpenetration of Civilizations* (Baltimore, MD: Johns Hopkins University Press, 1978).

a yearly Festival of the Orange: Other citrus production areas in Brazil also host a yearly orange festival. For instance, to give an example far removed from RS, the city of Anori, in the state of Amazonas, has one such festival. It produces, under its own admission, "rustic oranges." Anori is about 160 kilometers from Manaus, the capital of the state, in a region named Coari. An orange festival has been held there more than a dozen times already, during the first half of June.

the production of citrus in Brazil: G. Hasse, *A laranja no Brasil 1500–1987: A historia da agroindustria citrica brasileira, dos quintais coloniais as fabricas exportadoradas de suco do seculo XX* (São Paulo: Duprat & Iobe, 1987).

immigrants from Germany: G. Seyferth, *Nacionalismo e Identidade Etnica* (Florianópolois: Fundação Catarinense de Cultura, 1982); G. Seyferth, "Imigração e colonização alemã no Brasil: Uma revista da bibliografia," *Boletim Informativo e Bibliografico de Ciencias Sociais* 25 (1988): 3–55; G. Seyferth, *Imigraçao e Cultura no Brasil* (Brasilia: Editora da Universidade de Brasilia, 1990); G. Seyferth, "Identidade etnica, assimilação e citadiana: A immigração alemã e o Estado Brasileiro," *Revista brasileira de ciencias sociais* 9, no. 26 (1994): 103–22. Another scholar on the same topic of German immigration in Brazil, considered over the Braudélian *longue durée*, is C. Fouquet, *O imigrante alemão e seus descendentes no Brasil: 1808–1824–1974* (São Paulo & São Leopoldo: Instituto Hans Staden & Federação dos Centros Culturais, 25 de Julho, 1974).

On the more specialized topic of the installation of German immigrants in Rio Grande do Sul, see M. Mulhall, *O Rio Grande do Sul e suas colonias alemães* (Pôrto Alegre, RS: Bels/Instituto Estadual do Livro, 1974); T. L. Muller, *Colonia alemã/160 anos de historia* (Caxias do Sul & Pôrto Alegre, RS: Editora da UCS & Escola Superior de Teologia São Lourenço de Brindes, 1984); and J. Roche, *A colonizaçao alemã e o Rio Grande do sul* (Pôrto Alegre, RS: Globo, 1969). On German immigration in southern Brazilian locations other than Rio Grande do Sul, see, for instance, W. Aulich, *Parana e os alemães/Parana und die deutschen* (Curitiba: Grupo etnico Germanico do Paraná/Deutschethnische Gruppe von Paraná, 1953). On the German language in Brazil, a fine study is O. Nixdorf, "Die deutsche Sprache in Brasilien," *Südamerika* (Buenos Aires) 12, no. 2 (1961): 4 S. An excellent article on the sportive spirit within the German colonies in Rio Grande do Sul is Gilmar Mascarenhas de Jesus, "Imigrantes desportistas: Os Alemães no sul do Brasil," *Scripta Nova* 94, no. 108 (2001), available at http://www.ub.es/geocrit/sn-94-108.htm.|

Old Norse mythology: H. R. E. Davidson, *Gods and Myths of Northern Europe* (Harmondsworth, Middlesex: Penguin Books, 1964). Regarding the golden apples of Idun and their magical use for rejuvenation, there is much discussion as to whether it might be a late addition from Irish or Greek myths. In an early form of the tale of Loki and the eagle, in the *Haustlong*, composed by Thjodolf of Hvin, court poet to Harold Fairhair, who lived ca. 860 AD, Idun's absence causes the Æsir to age, with no mention of the golden apples.

CHAPTER 10

The ruling family in the Netherlands: They took the name of Orange and Nassau by combining the names of two principalities, in Southern France and in Germany, respectively. W. Ruizendaal, *Nassau & Oranje: 600 jaar geschiedenis van ons vorstenhuis van Engelbert I tot Willem-Alexander* (Baarn: Tirion, 1995). The former name derives from the Celtic word *arantio*, in which the prefix *ar* means "flat valley" and the suffix *sione* means a place, so, literally "the place of the flat valley." Pliny mentions the Roman name, Aransio. The visual pun of choosing orange for its emblematic color was thus a natural step for the royal family when the names of the city and of the fruit became homonyms.

Princeton University alumni: The university did not have orange and black as colors from its founding in 1746 as the College of New Jersey. When in 1756 it moved into Nassau Hall, at Princeton, this gave it a symbolic link to the

Dutch royal family. R. D. Smith, *Princeton* (Dover, NH: Arcadia Tempus Publishing Group, Inc., 1997). Only in the 1860s did students start bedecking themselves at athletic events with orange ribbons—probably in reference to William III, Prince of Orange, for whom Nassau Hall had been named. When they began to write class numerals in black on their orange ribbons, the two colors became paired. On April 5, 1866, George K. Ward (Princeton class of 1869) commented during a class meeting upon the lack of distinctive colors for the college. He then started agitating for the adoption of orange (with black printing) as the Princeton color, which was finally made official in 1868, at the time James McCosh was inaugurated as president. Gradually, orange and black together became the colors for Princeton. A. Leitch, *A Princeton Companion* (Princeton, NJ: Princeton University Press, 1978). When I taught at Princeton, in the 1960s, Hubert N. Alyea, who was a Princeton alumnus, taught general chemistry to freshmen. He was a most genial man. Whenever you would come across him in the hall, he would be whistling a little tune. Each year, for his last class, he would put on a show of experimental demonstrations. He was such a superb lecturer, and so imaginative with his experimental displays, that the largest lecture hall in the chemistry building overflowed with both students and faculty. At the very end of the lecture, Professor Alyea would show in a flask the colors of all the schools in the Ivy League in succession, singing a line or two from the corresponding anthems! The act was capped with his bellowing the Princeton song, "Old Nassau," while displaying orange and black in the flask. The orange came from mercuric iodide precipitate, and the black from the dark blue iodine-starch complex, iodine having generated from the iodide-iodate reaction with ions left over from the mercuric iodide–forming reaction (sodium hydrogen sulfite reducing iodate into iodide, in the presence of mercuric ions). H. N. Alyea, "Tested Demonstrations in General Chemistry," *Journal of Chemical Education* 32 (1955): 9; H. N. Alyea, chap. 1 in G. L. Gilbert et al., *Tested Demonstrations in Chemistry* (Granville, OH: Denison University, 1994), 1–49. The visual pun was taken one step further when Chaffey High School in Ontario, California (named for George Chaffey, who founded the city of Ontario, in the citrus belt; see chap. 5), adopted orange and black as its school colors, together with a school song copied from Princeton University's.

Pip is short for "pippin": O. Bloch and W. von Wartburg, *Dictionnaire étymologique de la langue française* (Paris: Presses universitaires de France, 1975); entry for "*pépin*," 475.

a nonfunctioning car: Many U.S. states passed "lemon laws" for car buyers' protection; see, for instance, http://www.CaliforniaLemonLawInformation.com.

a jingoistic speech [by] Sir Eric Geddes: *The Oxford Dictionary of Quotations*, 3rd ed. (Oxford: Oxford University Press, 1979), 223, no. 32; quoting the *Cambridge Daily News*, 10 December 1918, 3/2. While the verb "to squeeze" conveys predominantly the concrete act, its synonym, "to express," carries both literal and figurative meanings. As Francis Ponge so memorably wrote: "As in the sponge there is in the orange an aspiration to regain face after undergoing the ordeal of expression." F. Ponge, "L'orange," in *Le parti-pris des choses* (Paris: Gallimard, 1942). Ponge's beautiful way of putting it set a standard—one that I have striven to uphold in this book.

The autumn leaves owe their vivid colors: The autumn hues of foliage are the result of carotenoid and anthocyanin pigments predominating over green chlorophylls. The normal role of orange-colored carotenes in tree leaves is to transfer to chlorophylls a fraction of the energy they absorb from sunlight. Anthocyanins are the molecules responsible for some leaves turning a bright red. They are dissolved in the cells' aqueous medium (sap) rather than attached to the membranes of the disk-shaped chloroplasts, which, inside the leaf cells, store the chlorophylls and the carotenes. The color of anthocyanins depends on the acidity of the medium they bathe in (think of hydrangeas); they turn bright red when the aqueous solution is acidic.

In autumn, a corky membrane grows between the branch and the leaf stem, interfering with the flow of nutrients, produced by photosynthesis, from the leaf to the tree. This change switches off production of green chlorophyll, allowing the other pigments to dominate. Scientists do not yet know the "purpose" of the change in the color of leaves with the arrival of cooler weather and shorter days: G. Brumfiel, "The Warm Hues of Fall Foliage," *Scientific American*, October 15, 2001, available at http://www.sciam.com/article.cfm?articleID=0005DF26-7ABD-1D94-9275809EC5880105. Some believe that the red hues act as a sunscreen, which protects the leaves while the tree reabsorbs their nutrients: W. A. Hoch, E. L. Zeldin, and B. H. McCown, "Physiological Significance of Anthocyanins during Autumnal Leaf Senescence," *Tree Physiology* 21 (2001): 1–8. This is the belief of Kevin S. Gould, who points to the huge metabolic expenditure needed for a leaf to manufacture the red anthocyanins just before being shed by the tree: S. Neill and K. S. Gould, "Optical Properties of Leaves in Relation to Anthocyanin Distribution and Concentration," *Canadian Journal of Botany* 77 (1999): 1777–1782; K. S. Gould, K. R. Markham, R. G. Smith, and J. J. Goris, "Functional Role of Anthocyanins in the Leaves of *Quintinia serrata* A Cunn.," *Journal of Experimental Botany* 51 (2000): 1107–15. Other botanists believe that the function of the red anthocyanins is to protect the plant from drought, helping it to retain water: L. Chalker-Scott and L. H. Fuchigami, "The Role of Phenolic Compounds in Plant Stress Responses," in *Low Temperature Stress Physiology in Crops*, ed. P. H. Li (Boca Raton, FL: CRC Press, 1989), 67–79; L. Chalker-Scott, "Environmental Significance of Anthocyanins in Plant Stress Responses," *Photochemistry and Photobiology* 70 (1999): 1–9; L. Chalker-Scott, "Do Anthocyanins Function as Osmoregulators in Leaf Tissues?" in *Why Leaves Turn Red: Anthocyanins in Leaves*, ed. K. S. Gould and D. W. Lee, Advances in Botanical Research (London: Academic Press, 2002). Quite a few other explanations have been offered, including insect repellency, animal attraction, frost protection, and free radical trapping. This spread of opinions has two lessons: First, turning to scientists for answers to questions posed by nature, although it is our spontaneous inclination, is bound to be frustrating, since science studies not natural, but rather laboratory phenomena. Second, the notion of a consensus is alien to the scientific spirit; controversy is the normal state of science, which thrives on it.

cor-de-laranja: On naming colors, see L. Wittgenstein, *Remarks on Colour* (Berkeley, CA: University of California Press, 1990). And, regarding the color orange, Lamartine Babo's (1904–1963) "Hymn of the Brazilian Carnival," which salutes beautiful girls of every color thronging the streets of Rio, has the line "*Loiras, cor de laranja, cem mil . . .* " for its chorus ("blondes, orange-colored, one hundred thousand of 'em").

Many would be seen racing: Traffic was extremely dangerous. The popular singer Francisco (Chico) Alves was killed in a traffic accident on September 28, 1952. The whole city went into mourning; the public desolation can only be compared to that following the loss to Uruguay during the final game of the World Cup in 1950 (see chap. 9) or, outside Brazil, to that of Paris on the day another singer, Edith Piaf, was buried. On traffic lights, see W. Garrison, *Why Didn't I Think of That* (Englewood Cliffs, NJ: Prentice Hall, 1977); Garnet Nelson Jackson, *Garrett Morgan, Inventor* (Cleveland, OH: Modern Curriculum Press, 1993). Colored light signals on railroads were first used (five lights, later reduced to three) in 1915 on the Pennsylvania Railroad; see G. Wilson, *The Old Telegraphs* (London: Phillimore, 1976). Their introduction for transmission of information to the engineer may have been delayed by Lt. Col. Charles W. Pasley's statement, when he recommended use of lights at night, that "the colour of a luminous point or line cannot be distinguished at any distance": C. W. Pasley, *Observations on Nocturnal Signals in General* (Chatham, UK: Royal Engineers, 1823); see also C. W. Pasley, LtCol RE, FRS. *Description of the Universal Telegraph for Day and Night Signals* (London: 1823). Pasley was one of the leading military engineers of the nineteenth century.

Streetcars, known as *bondes*: The first electric streetcar in South America started running in Rio de Janeiro in 1892. In the 1950s, when I lived there, the *bondes* started suffering from competition with automobiles due to increases in both living standards and population (by 50 percent during the decade). Cutbacks were made. During the sixties, trolleybuses gradually replaced the *bondes*. Only the one to Santa Teresa was left running, as a tourist attraction, after 1968.

The *Lycée français* of Rio: Named Lycée Molière since 1982, it remains in the Laranjeiras district, but not in the identical location where I attended school.

Oranges groves on the outskirts: Hasse, *A laranja no Brasil*.

CHAPTER 11

Goethe: Goethe, *Selected Verse*, ed. David Luke (Harmondsworth: Penguin, 1964); *Goethe's Werke: Vollständige Ausgabe letzter Hand* (Stuttgart and

Tübingen: Cotta, 1827–1830). For critical comments, see Barker Fairley, *Goethe As Revealed in His Poetry* (London: J. M. Dent, 1932); E. M. Wilkinson, "Goethe's Poetry," in *Goethe: Poet and Thinker*, ed. E. M. Wilkinson and L. A. Willoughby (London: Edward Arnold, 1962), 20–34; David Wellbery, *The Specular Moment: Goethe's Early Lyric and the Beginnings of Romanticism*, Stanford, CA: Stanford University Press, 1996.

His poem "The Island": The quotation is from stanza 8 in *The Work of Lord Byron*, ed. Ernest Hartley Coleridge (London: Murray, 1904).

Kipling's "Rhyme of the Three Captains": from *The Writings in Prose and Verse of Rudyard Kipling*, vol. 11, *Verses 1889–1896* (New York: Charles Scribner's Sons, 1897).

The Spanish poet Antonio Machado: from *Poesías Completas*, ed. Oreste Macri and Gaetano Chiappini (Madrid: Espasa-Calpe and Fundación Antonio Machado, 1988).

The contemporary South African poet Don Maclennan: from *The Poetry Lesson* (Cape Town: Snailpress, 1995).

Federico Garcia Lorca: Federico Garcia Lorca, *Collected Poems*, ed. C. Maurer (New York: Farrar Straus Giroux, 1991). For his symbolism, see R. C. Allen, *The Symbolic World of Federico Garcia Lorca* (Albuquerque, University of New Mexico Press, 1972). A useful biographical study is that by Reed Anderson, *Federico García Lorca* (London: Macmillan, 1984).

the Greek poet Odysseas Elytis: Geoffrey Carson and Nikos Sarris, trans., *The Collected Poems of Odysseus Elytis* (Baltimore, MD: Johns Hopkins University Press, 2004 [1997]).

a proverb of Joyce's invention: James Joyce, *Finnegans Wake*, 2.1.253, available at http://www.trentu.ca/faculty/jjoyce/.

the poet of citrus par excellence: Numerous American poets other than Wallace Stevens might be construed as contenders for this honor. To mention only two among our contemporaries, Pulitzer Prize and National Book Award winner Mary Oliver and Gary Soto, the Young People's Ambassador for California Legal Assistance and the United Farm Workers of America, have both written poems on the orange. These poems are on their way to becoming, perhaps for their simplicity, high school favorites: M. Oliver, "Oranges," *Shenandoah* |4 (2001), 30; G. Soto, "Oranges," *Poetry*, June 1983. The latter, in the voice of a twelve-year-old boy, is about first love. The final image bursts with the incandescent joy of the new feeling, "I peeled my orange / That was so bright against / The grey of December / That, from some distance, / Someone might have thought / I was making a fire in my hands." It has appeared in the anthology *A Writer's Country: A Collection of Fiction and Poetry*, edited by J. Knorr and T. Schell (Upper Saddle River, NJ: Prentice-Hall, 2001).

"Sunday Morning": from *Harmonium* (New York: Knopf, 1923), 100–106.

the gods of paganism: for Stevens, "one of the first substitutes for Christianity," according to J. McCann, *Wallace Stevens Revisited: "The Celestial Possible"* (New York: Twayne Publishers, 1995).

religious promises of heaven: "Bereft of the supernatural, man does not lie down
paralyzed in despair." J. H. Miller, *Poets of Reality: Six Twentieth-Century
Writers* (Cambridge MA: The Belknap Press of Harvard University Press,
1966).

a paean to life: M. E. Brown comments that "even the chant of the ring of supple
and turbulent men, expressing their boisterous devotion to the sun, . . . is
dependent on their mutual sense of frailty": *Wallace Stevens: The Poem as Act*
(Detroit, MI: Wayne State University Press, 1970).

an original musical composition by Paul Hindemith: a piece for trumpet and
percussion.

two scientific presentations: Twenty years after this jubilee, Max Delbrück,
in his Nobel Prize acceptance speech, noted that the scientists' enjoyment of the concert and the reading had not been reciprocated,
only other scientists having attended their lectures: Max Delbrück,
"Nobel Lecture: December 10, 1969: A Physicist's Renewed Look at
Biology—Thirty Years Later," The Nobel Foundation, "Nobelprize.org,"
http://www.nobel.se/medicine/laureates/1969/delbruck-lecture.html.

"An Ordinary Evening in New Haven": from *Transactions of the Connecticut
Academy of Arts and Sciences* 38 (1949): 161–72; also in *Selected Poems*
(London: Faber and Fabler, 1993), 118–26.

the poet later expanded on it: The original version, selected for the public reading
in 1949, consisted of eleven stanzas, each subdivided into six segments of
three lines. The final version, first published in W. Stevens, *The Auroras of Autumn* (New York: Knopf, 1950) is composed of thirty-one stanzas, of which I,
VI, IX, XI, XII, XVI, XXII, XXVIII, XXX, XXXI, and XXIX (in that sequence)
are those of the earlier poem.

both of whom Stevens admired: A. W. Litz, "Particles of Order: The Unpub-
lished *Adagia*," in *Wallace Stevens: A Celebration*, ed. F. Doggett and R. Buttel
(Princeton, NJ: Princeton University Press, 1980), 57–77.

an oracular tone: J. N. Riddel, *The Clairvoyant Eye: The Poetry and Poetics of Wallace
Stevens* (Baton Rouge, LA: Louisiana State University Press, 1965), 256.

muffled repetitions: J. L. Borges, *This Craft of Verse*, The Charles Eliot Norton Lectures 1967–1968 (Cambridge, MA: Harvard University Press, 2000) 30–31,
shows beautifully and most convincingly how the doubling of identical
lines, using the example of a poem by Robert Frost, can convey two quite
distinct meanings.

plays with words: S. B. Weston, in *Wallace Stevens: An Introduction to the Poetry*
(New York: Columbia University Press, 1977), devotes her chapter 2 to
Stevens's joyful celebration of language, his use of puns, "to dramatize his
own unresolved feelings about the nature of reality" (10). J. Baird, *The Dome
and the Rock: Structure in the Poetry of Wallace Stevens* (Baltimore, MD: Johns
Hopkins University Press, 1968), 40, points to the meeting of minds with
Shelley, who had set his poem "Epipsychidion" in an island of the Aegean
"heavy with the scent of lemon-flowers."

Barry Lopez is wont to depict: B. Lopez, *Arctic Dreams* (New York: Bantam, 1996), for instance, 75.

CHAPTER 12

the "Dame à la Licorne" tapestries: A. L. Kendrick, "Quelques remarques sur la Dame à la licorne du Musée de Cluny (allégorie des cinq sens?)," in *Actes du congrès d'histoire de l'art* (Paris), no. 3 (1921): 662–66; C. Nordenfalk, "Qui a commandé les tapisseries dites de la Dame à la Licorne?" *Revue de l'art* 55 (1982): 53–56; C. Nordenfalk, "The Five Senses in Late Medieval and Renaissance Art," *Journal of the Warburg and Courtauld Institutes* 48 (1985): 1–22; M. B. Freeman, *La chasse à la licorne, prestigieuse tenture française des Cloisters* (Lausanne: Edita, 1983); A. S. Cavallo, *The Unicorn Tapestries at the Metropolitan Museum of Art* (New York: The Metropolitan Museum of Art, 1998); J.-P. Boudet, "Jean Gerson et la Dame à la Licorne," in *Religion et société urbaine au Moyen Age: Etudes offertes à Jean-Louis Biget* (Paris: Publications de la Sorbonne, 2000), 551–63; V. Huchard and P. Bourgain, *Le jardin médiéval: Un musée imaginaire* (Paris: Presses universitaires de France, 2002).

the Dutch still life: Svetlana Alpers, *The Art of Describing: Dutch Art in the Seventeenth Century* (Chicago: University of Chicago Press, 1983); S. Slive, *Dutch Painting 1600–1800* (New Haven, CT: Yale University Press, 1999); F. Lewis, *A Dictionary of Dutch and Flemish Flower, Fruit and Still-Life Painters, 15th to 19th Century* (Leigh-on-Sea, England: F. Lewis, 1973); I. Bergström, *Dutch Still-Life Painting in the Seventeenth Century* (New York: Hacker Art Books, 1983 [1956]); Arthur K. Wheelock, Jr., "Still Life: Its Visual Appeal and Theoretical Status in the Seventeenth Century," in *Still Life of the Golden Age: Northern European Paintings from the Heinz Family Collection* (Boston & Washington, DC: National Gallery of Art & Museum of Fine Arts, 1989), 14–15; A. Wallert, ed., *Still Lifes: Techniques and Style: An Examination of Paintings from the Rijksmuseum* (Seattle, WA: University of Washington Press, 2000); S. D. Muller, "Jan Steen's Burgher of Delft and His Daughter: A Painting and Politics in Seventeenth-Century Holland," *Art History* 12, no. 3 (1989): 268–98; G. Riley, *The Dutch Table: Gastronomy in the Golden Age of the Netherlands* (San Francisco: Pomegranate, 1994); M. A. Doty, *Still Life with Oysters and Lemon* (Boston: Beacon Press, 2001).

According to Sandrart: Joachim von Sandrart, *Teutsche Academie der Bau-, Bild- und Mahlerey-Künste* (Nürnberg, Germany, 1675).

the Norton Simon Museum: This world-renowned museum stems from the private collection of Norton Simon. The painting by Zurbarán was the wedding gift lavished by Simon on the Hollywood actress Jennifer Jones in 1971. It is said that when she came into a room, in a hotel for instance, she would take greater pleasure in seeing a bowl of lemons than flowers in a vase.

It shows, left to right: The notion that it is meant to be read, textlike, is compelling. The canvas is an elongated horizontal rectangle, measuring 60 × 107 cm, for a ratio of about 1.80, significantly greater than the Golden Section of 1.618. Its size makes it obvious that it was meant for hanging in a home.

a metal platter: Whether silver or pewter has not been satisfactorily resolved. The well-polished surface is shown as mirrorlike, reinforcing the illusion of the real.

a delicate cup: Why such a delicate, luxurious container rather than a glass or a metallic goblet? This may well be the key to the symbolic dedication of the painting to the Virgin Mary, as the whiteness of the china cup, together with its water content, symbolizes purity. But what is it made of? From my careful inspection of the painting in Pasadena, I decided against porcelain (china) and in favor of a light gray stoneware.

more a stand than a table: Some authors, such as W. B. Jordan, in *An Eye on Nature: Spanish Still-Life Paintings from Sanchez Cotan to Goya* (London: Mathiesen Fine Art, 1997) explicitly mention a tabletop; but it looks more to me like an altar top.

completed in 1633: The canvas is signed and dated. For Zurbarán's biography, a standard source is J. Brown, *Francisco de Zurbarán* (New York: H. N. Abrams, 1974).

Considerable work and forethought: Jordan, *An Eye on Nature*. The preparatory painting is a family property of the Clarks (acquired by Sir Kenneth Clark, the art historian who wrote *Civilization* and also had a television series by the same title), and was described by Alan Clark in his memoirs.

originally flanked by a platter of candied sweet potatoes: One may surmise that it was painted over and removed because the sweet potatoes were out of kilter. Zurbarán had placed this dish of delicacies in the foreground, in between the citrons and the oranges. In the final version of the painting, all the objects in it are aligned, left to right.

the arabesque of a leaf: C. Sterling, *Still Life Painting* (Paris: Editions Pierre Tisné and Universe Books, New York, 1959).

a symbolic homage to the Virgin Mary: The best source remains here J. Gallego, *Vision et symboles dans la peinture espagnole du siècle d'or* (Paris: 1968). One may also detect in the collection of objects chosen by Zurbarán a faint, lingering echo of the Jewish Feast of the Tabernacles.

the enduring concept of romantic love: see Denis de Rougemont, *L'Amour et l'Occident* (Paris: Plon, 1939).

recapitulates Western painting: Matisse's "Notes from a Painter" (1909), in *Henri Matisse: Ecrits et propos sur l'art*, ed. D. Fourcade (Paris: Hermann, 1972). Oranges were one of his favorite themes. There is a world, a chasm, between the 1912 "Nature morte aux oranges," which Picasso owned, perhaps for its elaborate classical technique, and the 1916 painting described here.

Malraux has argued: André Malraux, *Saturne: Essai sur Goya* (Paris: Gallimard-La Pléiade, 1950).

citrus crate labels: There are many guides and price lists addressed to collectors. Even a short bibliography ought to include T. P. Jacobsen, *Pat Jacobsen's*

Collector's Guide to Fruit Crate Labels (Pleasant Hill, CA: Crate Expectations/Patco
Enterprises, 1994); T. P. Jacobsen, *Pat Jacobsen's Millennium Guide to Fruit
Crate Labels* (Weimar, CA: Patco Enterprises, 2000); G. T. McClelland and
J. T. Last, *Fruit Box Labels: A Collector's Guide* (Santa Ana, CA: Hillcrest Press,
1983); G. T. McClelland and J. T. Last, *California Orange Box Labels: An Illus-
trated History* (Santa Ana, CA: Hillcrest Press, 1985); G. T. McClelland and
J. T. Last, *An Illustrated Price Guide to Citrus Labels* (Santa Ana, CA: Hillcrest
Press, 1995); K. M. Yee, *A Package Deal: The Art of Agriculture* (Salinas, CA: The
National Steinbeck Center, 2000); J. Chicone and B. Burnette, *Florida Cit-
rus Crate Labels: An Illustrated History* (West Palm Beach, FL: Brenda Eubanks
Burnette, n.d.).

CHAPTER 13

Proust's magnum opus: See André Aciman, "Proust's Way? An Exchange with Jan
van Rij, Lydia Davis, Marcel Muller, and Christopher Prendergast," *New York
Review of Books*, April 6, 2006, 70–72.
they were colored yellow: Why is it that the vocabulary descriptive of colors gradually becomes impoverished? Alternative names for the color orange, such
as amber or apricot, fall into disuse. Likewise, the word "yellow" tends to
smother other words. Sometimes they freeze into a stereotype from which
they no longer can be rescued, as in the "saffron robes of Buddhist monks."
And yet the entry for yellowness (436) in *Roget's Thesaurus* lists no fewer
than twenty-one adjectives besides "yellow": "citron, gold, golden, aureate,
citrine, fallow, fulvous, saffron, croceate, lemon, sulphur, amber, straw-
coloured, sandy, lurid, Claude-tint, luteous, primrose-coloured, cream-
coloured, buff, chrome." P. Roget, *Thesaurus of English Words and Phrases*
(London: J. M. Dent & Sons, 1937 [1912/1852]). "Lurid," for instance, has
maintained only its figurative meaning.
Candied Citrus Strips: The ancient recipe is from T. Dawson, *The good huswifes
Jewell, part 1* (London: Edward White, 1596). Another sixteenth-century
recipe, reproduced by P. Brears in *Banquetting Stuffe: The Fare and Social Background of the Tudor and Stuart Banquet*, ed. C. A. Wilson, First Leeds Sympo-
sium on Food History and Traditions (Edinburgh, Scotland: Edinburgh University Press, 1991), is for "sucade of peeles of Lemmons" (John Partridge,
The Treasurie of Commodious Conceits, and Hidden Secrets, 1573). Here is a
modernized version:

2 lemons (or oranges)

2 tablespoons rosewater

14 oz sugar

Halve the lemons, squeeze out the juice, cut the rinds into quarters, and scrape
out any remaining pith. Boil the rinds in one pint of water for 30 minutes,

changing the water three times during this period so that no bitterness remains and they become very tender. Make a syrup with the sugar, rosewater, and three-quarters of a pint (425 mL) of water from the last boiling, and simmer the peels in this mixture until they are translucent and the syrup has become as thick as thin honey. Store in sterilized jars until required.

A good address for outstanding candied lemon, and for either slices or segments of candied oranges covered with chocolate, is confiserie Blondel, rue du Bourg, Lausanne, Switzerland. The business was started in 1856 by Mr Blondel, a Vaudois gentleman. Their products are exquisite, truly mouthwatering. The last time I visited the shop, candied lemon was selling for eight Swiss francs a hundred grams.

orange wine: I am indebted to Madame Monique Gras for this traditional recipe from the French Southwest, which originated in the city of Castres with a cousin of Madame Gras. It makes a very nice aperitif.

Orange wine can be much more than a drink; it acts upon me as a Proustian reminiscence of an adorable and memorable lady, whose life might initiate a novella.

I shall call her Madame Gaillard. She lived in a small town, in Castres we might say. Let us leave it that way. Madame Gaillard was an elderly lady, of undepleted mental energy. She was in poor health and of failing eyesight, with cataracts clouding both eyes—which alone would set the time of my story, the 1950s in rural France. But she maintained an attitude toward people that was both amused and benevolent. She was constantly amused by people's foibles and actions. She would giggle at such stories. I can still hear the ripples of her laughter. She made you feel the young girl in her. This was so much nicer than to have to withstand the peevish matron one might have expected, whom she might have become. Oh yes, she did love gossip! It had become her main entertainment. She never watched television. She did not have a musical ear, and was not much of a radio listener. She could not drive a car, she had always refused to learn. The infirmities of old age prevented her leaving the house; she walked slowly and with difficulty.

She did not live by herself. Her son, Jean, remained a bachelor. He was in his mid-sixties, a dried stick of an old guy. He shared the huge house with her, and he was totally unobtrusive. He had two interests in his life that I know of, wine and the stock market. Yes, they do have something in common: optimism and a belief in the future. Jean was such an enthusiast. Not that he was a drunkard. He loved to go around in his old decrepit car and visit wine growers, to sit for hours in their cool, dry cellars and sample vintages, comparing the latest with earlier ones, speculating on what this particular wine would age into.

And yes, he spent a lot of time playing the stock market. Few people did this in France in the 1950s. He did it wisely, mostly to amuse himself. He would invest little. He was in for the gamble. From what I heard, he was a consistent loser. But he had an unenterprising mentality, or was cautious enough to bleed in a trickle rather than in a flood the small fortune his father had left him. Madame Gaillard had only affectionate yet undisguised contempt for her son's doings. She was no longer amused by him.

There is a certain type of French townhouse that I can no longer see without it reminding me with a pang of Madame Gaillard's: built in the eighteenth century, sealed in upon itself with closed shutters, two-storied, large, with a big vaulted entrance closed with a heavy wooden double door, a gate wide enough for carriages to have entered the inner courtyard and stables. Madame Gaillard's house was out of a notary's diary. Not for its unremarkable plain architecture, but for the riches inside! A room-by-room description would be tedious; such an inventory might be of interest only to museum curators or dealers in antiques. But their blood would quicken, and their greed would become palpable.

Madame Gaillard's husband, about the time they were married, had inherited the family bank. They wed young. It was, rather unusual then, a marriage of love. They had money. They enjoyed leisure time. The husband—was he named Edmund, perhaps?—worked hard, but his work week was short. He wanted the free time. He was fortunate in having a manager he trusted, who ran the bank for him, both efficiently and honestly. Yes, I know, this sounds unreal—but this is how it was. I am telling you a true story, and I am sorry if it does not have a twist for the cynical in us to enjoy. It is definitely not a story out of Balzac or Maupassant.

The Gaillards took long weekends when the notion was still alien to the French. They took off for weeks at a time. They had a singular passion: they both communed with beauty, not in landscapes, but in man-made objects. They toured Europe as if they had been wealthy British aristocrats or heirs to a South American silver mine. This was in the 1920s. They had good taste. In art appreciation they were self-taught. But this was also in the aftermath of World War I. You could acquire treasures for a song. Heirlooms were sold by families shattered by the loss of their providers. Often, both the husband and the only son had been killed. The ensuing financial straits were dire indeed for such unfortunate families, of zero means. The Gaillards bought wisely; that is to say, they bought only what they both cared deeply for. They purchased antique furniture, paintings, and sculptures, beautiful objects of every description from past centuries. They traveled predominantly in France, but also a great deal in Switzerland, Austria, and northern Italy. Their house was that of collectors. Some of the pieces they owned were better than anything at Musée des Arts Decoratifs, Musée de Cluny, or the Malmaison. A few were worthy of the Louvre.

Was it leukemia? Monsieur Gaillard died when Jean was only ten, and Madame Gaillard became a widow at thirty-five. The bank was easily sold, as it had prospered. She and her son were now, if not wealthy, affluent enough not to have to work for a living. She started to live for her memories. The dozen years of married life had been so happy, so full, that she could make them into a lifetime of bedazzled recollection.

You know what a good teacher is like, telling compelling stories. There are also a few great teachers. They are uplifting. They are enthusiastic. They make what they know sound fascinating, to the point of seeding in some of their pupils a lifelong pursuit. Madame Gaillard had such charisma for two reasons: her passion for the genuinely beautiful and her gift for storytelling. To enter a conversation with her was such a joy. You would sit thrilled, carried along, and realize, suddenly, that she had been talking and that you had been listening avidly for a good two hours, even three hours. She would mention some object, a Limoges enamel perhaps. Her photographic memory would take her back to its dusty discovery in a brocanteur's heap, to the ridiculously low price she and her husband had paid for it. She would tell you which of the Limoges workshops it had come from, and she was able to date it rather accurately. She owned thus quite a few truly precious things, from the late Middle Ages and the Renaissance.

You might notice, on top of a dresser, an enameled charger with an embossed whirl motif. She would tell you that it was a fifteenth-century Venetian piece, that the Museum of the Hapsburgs in Vienna held a similar one, and that she and her husband had bought it in Lucca, on their way to attend an opera in Verona (which, she said with a chuckle, she did not care for very much).

Or she would bring out and show you a smallish bronze sculpture, no taller than eight inches. It was one of her favorites. She would proceed to reveal that it was thirteenth-century French. She believed it to have been cast around 1280 and to have been part of the treasure in one of the tall Gothic cathedrals completed at that time in northern France. On stylistic evidence, she thought it came from the chapter house in Laon. It showed a seated man, the head turned sideways, laughing his heart out at the silliness he had just heard: a unique instant and a whole mode of being, which the sculptor had snapped up.

You might ask Madame Gaillard about a small ivory plaque, among dozens of other beautiful and rare objects scattered in her sitting room. She would tell you it was from the tenth century—tenth century!—and that it came from Constantinople, probably as war loot for the Venetians at a later time. She and her husband had found it in an antique shop in Salzburg. Which brought up the memory of the rest of the purchase they had made from the same dealer on that day. She left the room, and after some time she brought back, from another of her many rooms, all bulging with dusty treasures, a lumpy mortar with flaring sides, from the Tyrol, which she dated from the sixteenth century, on the evidence of the angular handle.

And this would go on, as you sat transfixed at this evocation of past happiness, still alive for this woman, since yesterday was her life. Today was, for her, mostly the opportunity to talk about yesterday, about that magic

Golden Time of public peace and private travel, of marital joys, and of being with her husband the lone tourists in museumlike ancient cities from Strasbourg to Prague, from Lucerne to Sienna, and sometimes, farther north, to Bruges and Antwerp, The Hague and Amsterdam.

You could not part from her without receiving a small gift. She was very generous. She knew that giving away a single artifact would not even dent her considerable hoard. The last time I saw her, I went home with a lovely alms dish, depicting Adam and Eve by the Tree of Knowledge. But not before we had sipped together a few glasses of her *vin d'orange*.

The tale I have spun here is about fifteen percent fiction. Since I do not enjoy total recall for the *objets d'art* mentioned, I have leaned heavily on one of the most readable and entertaining chronicles by S. Melikian, *International Herald Tribune*, Paris, March 2–3, 2002, 16.

Cedar Key: F. Sargeant, *Frank Sargeant's Secret Spots, Tampa Bay to Cedar Key: Florida's Best Saltwater Fishing* (Lakeland, FL: Larsen's Outdoor Publishing, 1992); C. Kummer, "Citrus Preserved," *Atlantic Monthly* 4 (1997), 104–8.

The English word obviously derives from *marmelada*: Tania Styles, "Confections à la mode: Revising the OED 's Etymologies," *OED News*, December 2001, http://www.oed.com/public/news/0112_2.htm. The lexicon thus provides one more example of the tight connections that have existed for centuries among the two Atlantic countries, Britain and Portugal.

CHAPTER 14

orange mousse and feijoada: D. Botafogo, *The Art of Brazilian Cookery* (New York: Hippocrene Books, 1993 [1960]); E. T. Ang and M. R. Fabrizio, *Delightful Brazilian Cooking* (Seattle WA: Ambrosia Publications, 1993); J. Peterson, D. Peterson, and S. V. Medaris, *Eat Smart in Brazil: How to Decipher the Menu, Know the Market Foods and Embark on a Tasting Adventure* (Madison WI: Ginkgo Press, 1995); C. B. Pinto, *Brazilian Cooking* (Edison, NJ: Chartwell Books, 1998). An illustrated cookbook of Brazilian cuisine, which provides recipes modified to incorporate ingredients available in the United States, is C. Idone, *Brazil: A Cook's Tour* (New York: Christopher N. Potter, 1995).

The Ugli: Hodgson, "Horticultural Varieties of Citrus." The part of the story on the Ugli being a registered trademark is highly similar to the current controversy about taxol, a natural product, being turned into taxotere, a drug. See J. Goodman and V. Walsh, *The Story of Taxol: Nature and Politics in the Pursuit of an Anti-cancer Drug* (Cambridge: Cambridge University Press, 2001).

EPILOGUE

If I may paraphrase our great Tao philosopher Chuang Tzu: "Once I, Chuang Tzu, dreamed I was a butterfly and was happy as a butterfly. I was conscious that I was quite pleased with myself, but I did not know that I was Tzu. Suddenly I awoke, and there was I, visibly Tzu. I do not know whether it was Tzu dreaming that he was a butterfly or the butterfly dreaming that he was Tzu. Between Tzu and the butterfly there must be some distinction. [But one may be the other.] This is called the transformation of things." See, for instance, "Chuang Tzu's Butterfly," http://www.srds.co.uk/begin/chuang1.htm. See also Borges, *This Craft of Verse*, 29–30.

Index